singing grass, *burning sage*

DISCOVERING WASHINGTON'S SHRUB-STEPPE

A NATURE CONSERVANCY OF WASHINGTON BOOK

TEXT BY JACK NISBET

GRAPHIC ARTS CENTER PUBLISHING®

acknowledgments

While accepting responsibility for any errors that might have landed in the text, I would like to gratefully acknowledge the following professionals who so kindly shared their knowledge and enthusiasm. Peter Dunwiddie, Pamela McAllister, and Curt Soper of The Nature Conservancy and David G. Gordon provided valuable insights and attention. I would especially like to thank my wife Claire for her editorial assistance, and Emily and Jamie for setting up the tent.

Rich Bailey, Bureau of Land Management; Kathryn Beck, Calypso Consulting; Pam Camp, Bureau of Land Management; Florence Caplow, Calypso Consulting; Dennis Dauble, Pacific Northwest National Laboratory, Battelle; David Godlewski, Cominco American; Patti Ensor; Adeline Fredine, Colville Confederate Tribes; Michelle Gerber, Fluor-Daniel Hanford; Dave Geist, Pacific Northwest National Laboratory, Battelle; Lisa Hallock, Washington Department of Natural Resources; Steve Herrmann, Evergreen State College; Bill Leonard, Washington Department of Ecology; Rich Landers; Jack Linnville; Kirk Phillips; Bob Pyle; Bill Rickard, Battelle; Ann Sharley-Hubbard, Bureau of Land Management; Andy Stepniewski; John Withy; Richard Zack, Washington State University

PHOTOS:

Cover and title page: Saddle Mountain sagebrush and horned lark; Page 2: Bitterroot. 4: Lupine on Rattlesnake Mountains, Arid Lands Ecology Reserve. 5: Purple lupine. 117: Common blue butterflies. 118: Luzuli bunting. 120: Simpson's barrel cactus.

introduction

A CUT IN THE CANYON WALL

The welcome heat of an early summer afternoon lingered inside the coulee, dragging the fresh scent of sage through the air. Scattered wildflowers added spikes of purple and gold to the gray-green blur of bunchgrass and sagebrush. As I walked along the canyon floor, the music of hidden sparrows and insects quieted in advance of my footsteps, only to start again a short distance away. Both sides of the coulee were skirted with deep ruffles of fractured stone, and above these talus slopes rose thick bands of dark basalt palisades. All along their length, the cliffs were splotched with bright orange and yellow lichens.

A bunting's song rang from a high branch, and its chorus followed me up through open sagebrush to the very top of the escarpment.

LAZULI BUNTING

The western wall of the canyon was marked by a deep cleft that sliced through the layers of basalt, extending an invitation to climb. I crossed to the base of the talus and scrambled up an unstable sea of broken rocks and boulders until I met a faint animal trail that curled into the cleft. There I found myself in a cool, shady cavern about twenty feet across, like a large roofless room with one wall fallen away. In wetter seasons I would have been standing inside the last cascade of a tall waterfall, but now the floor was bone dry save for a shrinking pool of still, clear water.

The evening wind freshened outside, penetrating the depths of the cavern with a soft moan. Helped by the trunk of an aged serviceberry bush, I wriggled up the back wall of the cascade to a series of spillways that rose through the basalt cliffs like a giant staircase. I hoisted myself up several more steps of various heights to a small pocket valley whose moist bottom was dotted with small clumps of cattails and shimmering aspen trees. A bunting's song rang from a high branch, and its chorus followed me up through open sagebrush to the very top of the escarpment.

The wind that buffeted the edge of the cliffs swirled out over a panorama of sculpted stone. Far below, the floor of the coulee was filled with afternoon shadows, but up on top the sun's late rays still slanted across the landscape, igniting acres of sage with pure light. Patches of tall grass waved maroon and beige around a faraway barn. Beyond that single structure, the land rolled away in great camel humps toward a distant horizon, its smooth line cut by the jagged cone of a snow-clad volcano.

The coulee where I stood lies in the midst of the arid country of eastern Washington, an area known variously as the Columbia Basin and the Columbia Plateau; Coulee Country and Sagebrush Country; the Scablands, the Big Bend, and the Dry Side.

CURLY DOCK AND BUNCHGRASS ATOP THE WHITE BLUFFS, COLUMBIA RIVER

"They have only thin vegetation on them, —not enough to disturb or conceal the beautiful forms, the curves which the waves leave on the hills they deposit. Their colors are very subdued, pale salmon from the dead grass, or light green like a thin veil, with the red earth showing dimly through."

—TRAVELER CAROLINE C. LEIGHTON, 1866

"We went north through the coulee, its perpendicular walls flooring a vista like some old ruined, roofless hall, down which we traveled hour after hour."

—Lt. Thomas W. Symons, 1879

Palouse River Canyon

❧ *The dramatic landscapes of eastern Washington offer vivid testimony to the creative forces of tectonic thrust and volcanic uplift, and to the erosive powers of wind and water.*

"That long ride without water or wood where last summer we suffered so much from the dust & heat, on this occasion we were caught in a snowstorm, there was a high wind blowing which drove the snow into our eyes & drifted it across the track so that it was with great difficulty we found our road. . . . The wind which generally blows there hard enough to take the hair off one's head cut right to the bone almost stopping the circulation."

—LT. CHARLES WILSON, 1861

The land embraced by the long rugged "S" of the Columbia River forms the heart of this region, but its horizons sweep north up the Okanogan River and south beyond the Snake; east to the foothills of the Bitterroot and Blue Mountains and west to the steep wall of the Cascade Range. The characteristic elements of this territory are dry rolling plains sprinkled with sagebrush and bunchgrass. The summers here are hot and dry, the winters cold; there is enough precipitation to support grasses and shrubs, but not enough for continuous forest. Today, range scientists classify this particular combination of vegetation, topography, and climate as "shrub-steppe," a designation that evokes the boundless Eurasian plains of Anton Chekhov and Attila the Hun.

The vast spaces of this wide domain are full of wonder and surprise, as slight differences in elevation, rainfall, soils, exposure, and landforms create a rich mosaic of diverse habitats. Treeless breaks above The Dalles give way to groves of Garry oaks along the Klickitat. Scablands of weathered rock lap against the gentle symmetry of the Palouse Hills. Pothole lakes sparkle beside bare gravel bars, and wind-whipped sand dunes loom above crowded cattail marshes. Stark cathedral walls within the Grand Coulee hide caves layered with relics of ancient dwellers. Around the Columbia's Big Bend, massive ridges diminish into teardrop islands of rock and soil that float across the plateau like ghost ships on a rolling sea.

Over the course of the past two centuries, the grand tapestry of Columbia Basin shrub-steppe has been steadily unraveled. Of the ten and one-half million acres of continuous sagebrush and bunchgrass habitat encountered by early European explorers, almost two-thirds has disappeared entirely, and much of the rest has been irrevocably changed. Many species of plants, fish, and animals have been reduced to populations

Sagebrush steppe in winter

"I can hardly sit down to write,

not knowing what to gather first."

—BOTANIST DAVID DOUGLAS, 1826

SPRING WILDFLOWERS ON THE QUILOMENE WILDLIFE RECREATION AREA

both tiny and fragmented. The loss of so much of this original ecosystem only magnifies the value of what is left. The dramatic landscapes of eastern Washington offer vivid testimony to the creative forces of tectonic thrust and volcanic uplift, and to the erosive powers of wind and water. Its flora and fauna reflect the elegant adaptations of living things to change and to hardship. The clear and complex beauty of healthy shrub-steppe exemplifies the remarkable vitality of sustained existence. Yet many of the pieces of shrub-steppe that remain viable—places that still breathe with a sense of cohesive life—face an uncertain future. It is time to appreciate how precious these remnants of the original sagebrush country really are.

That, simply put, is the purpose of this book: to explore the many faces of the shrub-steppe, to probe its hidden niches and the specialized plants and creatures that inhabit them, to examine how native tribes lived, and to ponder the recollections of early visitors. It is a chance to look at the way civilization has played across the Columbia Basin, and to think about the way decisions of the past and present will affect its future. It is a visual celebration of a landscape, full of power and diversity, whose resiliency is being put to the test.

The purpose of this book is to explore the many faces of the shrub-steppe, to probe its hidden niches and the specialized plants and creatures that inhabit them, to examine how native tribes lived, and to ponder the recollections of early visitors.

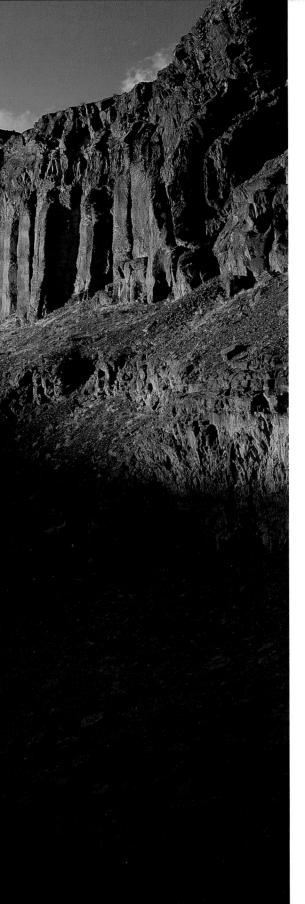

a fine field for the imagination

THE LANDSCAPE OF THE SHRUB-STEPPE

Since the beginning of recorded history, visitors to the Columbia Basin have reacted to its bare splendors with feelings that range from uneasiness and thirst to curiosity and unabashed awe. The journals of Meriwether Lewis and William Clark provide the first written descriptions of this part of the country, which they called the "Great Columbian Plain." On their journey downstream in the fall of 1805, Clark remarked mainly on practical matters such as the shortage of firewood and the abundance of rapids, but he did take time to marvel at two black boulders in the river—one that looked to him like an oversized hat, and another that resembled the hull of an enormous ship.

❦ **"These Lands are wholly composed of Strata of Rock from 10 to 30 ft thick . . . often like the flutes of an organ at a distance . . . The pillars are split also in various directions as if broken & cracked by a violent blow."**

—SURVEYOR DAVID THOMPSON, 1811

COLUMNAR BASALT WITH LICHENS, MOSES COULEE

◄◄ DUSTY LAKE   ◄ JUNIPER DUNES

Six years after the American expedition skirted the southern fringe of Washington's shrub-steppe, Canadian fur agent David Thompson canoed the entire course of the Columbia. As he floated around the Big Bend in July of 1811, he reveled in the wild scenery, the strong-scented shrubs, and the meadows that seemed to spring out of sheer basalt. He conjured whimsical images of castles and fluted organ pipes from the rock formations that lined the river, and found the landscape "in all its wildest forms, a fine field for the imagination to play in." British army officer Henry James Warre found his imagination completely overwhelmed when he stood at a vista overlooking Palouse Falls in the summer of 1846: "The number of rainbows & curious pointed basaltic Rocks; the spray & noise & the wildness of the boundless barren prairie above are such as can hardly be conveyed."

For many generations before the appearance of these early journal keepers, native peoples had been constructing their own interpretations for the geography of their homeland. A Yakama legend describes how Mount Hood, in a fit of jealousy, shattered the head of Mount Adams with a great war club, littering the path to a good fishing lake with broken stones. Another story tells how Warm Wind and Cold Wind struggled against each other, blowing up tornados and tidal waves, alternately freezing and thawing the ground around the White Bluffs section of the Columbia. The Spokanes recall a great flood that washed across the land, forcing their people to flee to the hills.

Several Basin tribes attributed the creation of river rapids and other significant landmarks to the legendary Coyote, or *Spill yay*. According to Sanpoil tradition, the Columbia once ran through the Grand Coulee, providing plenty of fish for the families who lived there. Then one day Coyote decided he would marry two of the local girls; when

Dry Falls at the head of Lower Grand Coulee

"The country was covered with a luxuriant growth of bunch-grass, with here and there a tract of sage-brush. The soil is of firm and excellent quality. Quite a large number of cattle were seen, all of which had to descend to the river for water."

—Thomas W. Symons

White Bluffs, Hanford Reach

❧ **"The men worked long and hard catching fish, while their women cut and cured it on the banks."**

—Colville-Okanagan author

Mourning Dove

Wishram woman preparing salmon, 1909

they both refused his proposal, he became so angry that he used his immense powers to change the course of the river, leaving the coulee dry and forever barren of fish.

The same powerful forces that were personified in these oral traditions also appeared in the writings of early naturalists. In 1835 the Reverend Samuel Parker declared that the dry floor of Grand Coulee "was undoubtedly the former channel of the river," setting off a century of theorizing as pioneer geologists puzzled over the origin of the Plateau's terrible beauty. Many hypotheses were advanced, and several episodes of a long saga were laid in place, but no one managed to twist all the strands of evidence into a coherent narrative.

Then in the 1920s, a geologist named J Harlen Bretz arrived on the scene. A tireless detective, Bretz probed wide river valleys that flowed only with sagebrush. He peered inside caves perched high above valley floors, and measured house-sized boulders plopped in the middle of wheat fields. He pondered strange ridges of gravel that wavered across the plains, and stood beneath dry waterfalls of incomprehensible scale. Over the course of five decades, Bretz and others he inspired used these clues to piece together the climax of the region's geologic story.

The first known chapters of this epic date back over a billion years, to a time when the slow workings of plate tectonics were adding land mass to the western edge of our continent. As recently as two hundred million years ago, ocean tides still lapped at what is now the Idaho Panhandle. Gradually more land rose from the sea, forming a shallow bowl in what is now eastern Washington and northeastern Oregon. A network of streams and lakes, the precursors of today's Columbia River, meandered across the gently sloping surface toward the Pacific.

Around seventeen million years ago, an enormous underground reservoir of viscous magma began bubbling up through wide fissures along the southeast margin of this basin. Following the land's natural inclination, pulses of molten rock flowed across the region, puddling against the base of the Okanogan Highlands to the north and pouring westward through rounded coastal foothills to hiss into the ocean. This eruption lasted only a short time, leaving a huge platter of magma that slowly cooled into hard brown basalt. Similar releases of lava were repeated more than three hundred times over the next ten million years, until layers of basalt ultimately spread over thousands of square miles, with accumulated depths of over eight thousand feet in some places.

As recently as two hundred million years ago, ocean tides still lapped at what is now the Idaho Panhandle.

The climate of that time was warm and humid . . . within this greenery wandered three-toed horses, tapirs, and peccaries; heavy-set rhinoceroses, long-jawed mastodons, and a horned gopher with a profile as unlikely as a jackalope's.

Sometimes tens of thousands of years passed between magma flows. During these interim periods, waterways would cut across the Plateau through fractures in the basalt, carrying sediments from the surrounding highlands. The climate of that time was warm and humid, and the accumulating soils nurtured lush forests that included maples, ginkgos, witchhazel, and bald cypress. Within this greenery wandered three-toed horses, tapirs, and peccaries; heavy-set rhinoceroses, long-jawed mastodons, and a horned gopher with a profile as unlikely as a jackalope's. Such flora and fauna would thrive until another pulse of magma flowed across the landscape, incinerating all the life in its path. In certain bogs and shallow lakes, the lava encased plants and animals that lay buried in mud, preserving relics such as the Blue Lake rhinoceros and the petrified tree trunks in the Ginkgo National Monument near Vantage. In some places, this pattern of leafy growth alternating with sudden destruction repeated itself several times over.

During this same Miocene period, tectonic pressure from the south slowly wrinkled up the Yakima Fold Belt, a series of east-west ridges that include Badger Mountain, Umtanum Ridge, the Rattlesnake Hills, and the Saddle Mountains. The ancestral Columbia River, which had been pushed to the northern and western margins of the plateau by successive basalt flows, cut its way south through the rising Saddle Mountains at Sentinel Gap. Like a band of gently waving sea kelp, the river curved eastward around the flank of Umtanum Ridge, bent back south to join the Snake in the Pasco Basin, then held a westerly course through the Horse Heaven Hills at Wallula Gap.

As the basalt flow epoch drew to a close around seven million years ago, grinding plates at the edge of North America continued to uplift the Cascade Range. While giant volcanoes showered the region with ash, the Columbia River maintained its channel

dams, water, and life

WOMAN DRIVING TEAM OF HORSES PULLING COMBINE, EARLY 1900S

For millenia, the arid climate of the Washington shrub-steppe determined which species would grow and thrive and which would wither and die. That monopoly over life and death was broken in the 1930s, when the first dams were built to regulate the flow of the Columbia River. The Columbia Basin Project made available huge amounts of water, distributing these across the parched land in predictable quantities.

Suddenly, the delicate balance of the aridlands ecology was irrevocably altered. Within a couple of decades, an area the size of Rhode Island was converted from shrub-steppe into productive orchards and farmland.

However, the governmental dream of a land of small, independent farmers was remarkably short-lived. Only the Grand Coulee Dam would actually supply water to an intricate web of irrigation canals. Instead of diverting water to irrigate family farms, subsequent dams were constructed to generate electricity for aluminum smelters and other industrial uses.

Less imposing than the dams, but of equal significance to the water balance of the shrub-steppe, are the thousands of wells that were drilled to provide irrigation water in recent decades. Today, the eternal cycles of aridity, heat, and sun across much of the Columbia Basin have been replaced by the endless rotations of central-pivot irrigation systems.

—*Peter W. Dunwiddie*

Mammoths, mastodons, and giant ground sloths feasted on shrubs and small trees, while herds of camels, antelope, and long-horned bison grazed on grasses. These herbivores were pursued by such outsized predators as saber-toothed tigers, American lions, short-faced bears, and dire wolves.

through the gradually rising mountains. Eventually the cordillera grew tall enough to snag weather systems blowing in off the Pacific, robbing moisture from the east side. Its lush forests died out and were replaced by a drier mix of pine and sagebrush, marking the advent of the shrub-steppe.

About two million years ago, a change in climate ushered in a long Ice Age. Tongues of great glaciers licked down across Canada, reaching far enough south to block the Columbia's channel through the Grand Coulee, altering the course of the big river yet again. South of the icefields, salmon plied meltwater streams. Mammoths, mastodons, and giant ground sloths feasted on shrubs and small trees, while herds of camels, antelope, and long-horned bison grazed on grasses. These herbivores were pursued by such outsized predators as saber-toothed tigers, American lions, short-faced bears, and dire wolves. Through cycles of changing weather the glaciers advanced and retreated, grinding the basalt surfaces into a fine soil known as loess (pronounced any number of ways, one of which rhymes with fuss). Prevailing winds distributed this rich loess across the Plateau in rounded dunes hundreds of feet deep.

Around twenty thousand years ago, one finger of this great ice sheet descended across the mouth of the Clark Fork River in northern Idaho. The lobe dammed the river, creating an immense body of water known as glacial Lake Missoula. When that lake reached a critical depth, it breached the ice dam, unleashing a flood of Biblical proportions. A gigantic wall of water rushed from northern Idaho across eastern Washington at speeds up to fifty miles an hour. The deluge overwhelmed existing watercourses, stripping away tons of topsoil and scouring coulees deep into ancient basalt. The force of the water carved huge waterfalls and gnawed out pothole lakes.

Frost-splintered basalt

"Passed a Pillar like rock split in pieces as by accident, in every horizontal direction."

—David Thompson

Icebergs carrying Montana boulders rammed into horizons of volcanic rock. Violent currents sculpted mesas shaped like steamboats and fancy hats. Swirling eddies laid down enormous islands of gravel, and standing waves left giant ripple marks undulating over the plains. When the flood reached the narrow bottleneck of Wallula Gap, water backed up into an enormous lake that filled the Pasco and Quincy Basins. The pressurized torrent shot through the gap, shearing the walls of the Columbia Gorge and flooding the lower Willamette Valley before finally expiring at sea.

Such a gargantuan flood, one of the largest ever measured on earth, happened not once but dozens of times, with the last one careening across the Columbia Basin around thirteen thousand years ago. J Harlen Bretz, who assembled the convincing evidence of these glacial floods, coined the term "channeled scablands" to describe the maze of channels scoured out by the water's rush. He described the scablands as "wounds only partially healed," and saw these lesions as vivid reminders of the cataclysmic events that gave the landscape of the Inland Northwest its special character. "The region is unique," Bretz wrote. "Let the observer take the wings of the morning to the uttermost parts of the earth; he will nowhere find its likeness."

"After a receding cataract has destroyed itself, what features of the gorge it made will record its former existence?"

—Geologist J Harlen Bretz, 1932

Palouse River Canyon from the air

bunchgrass man

HUMAN HISTORY

In the wake of the Lake Missoula floods, many drainages in the Columbia Basin lay battered and sterile. But where the torrential waters had not flowed, hardy vegetation still flourished, and it was not long before lichens and mosses crept onto the bare soil, and seeds of shrubs and grasses sprouted in the silt and rubble. Large mammals followed the retreating ice to graze in the new grass, and other animals colonized their own niches in the budding ecosystem. In creeks and rivers, beds of fine gravel laid down by the deluge provided ideal spawning grounds for salmon and trout. No one knows exactly at what point people entered this scene, or whether they were present

"Before the white people settled here, no one could destroy the Puh-tuh num (rock pictures). If you rubbed them all over with any kind of coloring, any kind of mud, the next morning they would be all bright and fresh as ever."

—CHIEF SLUSKIN TO L. V. McWHORTER, 1917

◄◄ HORSE HEAVEN HILLS ◄ BITTERROOT

ANCIENT ROCK ART FROM BIG EDDY UPSTREAM FROM THE DALLES

before the great floods, but scientific dates for the last flood lie tantalizingly close to the earliest recorded human relics in eastern Washington.

In 1987, workers laying irrigation pipe at the Richey apple orchard in East Wenatchee uncovered an orderly cache of stone tools. Some of the pieces were crafted of lustrous chalcedony, others of obsidian and quartz. Careful dating revealed that some of the spear points had been made and hafted over 11,000 years ago, in what is called the Clovis style; minute blood flakes recovered from their edges have been traced to animals from the bison, deer, and rabbit families. The beauty and size of these artifacts have led some researchers to believe that they must have been used only for ritual purposes, but archaeologist R.M. Gramly has theorized that the cache was a portable tool kit for hunting and butchering animals. In addition to points and blades, the site contained two sets of bone rods fashioned from the legbones of mammoths; Gramly thinks these may have been protective guards for the runners of a small sled that the hunters used to transport meat.

Numerous other discoveries establish that humans were abroad on the Columbia Plateau from at least this time forward. Radiocarbon tests from the Marmes rock shelter near the mouth of the Palouse River indicate intermittent habitation from ten thousand to about four thousand years before the present. Animal bones associated with the camp include fox, deer, bison, elk, pronghorn antelope, and salmon. The almost complete human skeleton dubbed Kennewick Man, which washed out of a gravel bank at the confluence of the Snake and the Columbia Rivers in 1996, has been reckoned at over nine thousand years old. A site in Lind Coulee, dating back to about eight thousand years ago, has yielded stone tools and bones of the extinct long-horned bison, as well as remains of muskrat, beaver, goose, and teal.

The almost complete human skeleton dubbed Kennewick Man, which washed out of a gravel bank at the confluence of the Snake and the Columbia Rivers in 1996, has been reckoned at over nine thousand years old.

KLIKITAT WATERTIGHT BASKET

OKANAGAN WOMAN GATHERING HUCKLEBERRIES, 1916

Pollen records indicate that climatic conditions during this postglacial era gradually cycled into a warm, dry period that precipitated changes in the vegetation, and over a relatively short period of time most of the large animals faded from the scene. Mammoths, early horses, and camels, along with such predators as short-faced bears and saber-toothed tigers, all apparently disappeared between eleven and ten millenia ago.

Any human role in these extinctions remains a subject of debate, but people certainly adapted to the changes in climate more successfully than mammoths. While the exact lineage that threads between the owners of the Clovis butchering kit and modern Native Americans is uncertain, the first written accounts of the Columbia Basin record bands of people all up and down the major rivers. These Plateau cultures fell into two distinct language groups. Interior Salish speakers, including tribes known today as Spokane, Sanpoil, Nespelem, Okanagan, Moses-Columbia, and Wenatchi, lived in the northern part of the Big Bend country. At Priest Rapids on the Columbia, the talk shifted to the Sahaptian languages of the groups now called Wanapum, Yakama, Klikitat, Walla Walla, Palus, Cayuse, Umatilla, and Nez Perce. These people called the Columbia the "Big River" and depended on it as a trade route and food source.

Tribes usually located their winter camps on the Columbia or its tributaries in order to take advantage of milder weather, wintering animals, and available fish. With the first appearance of greenery, they traveled inland to gather shoots and bulbs of the lomatiums known as desert parsleys or biscuitroots. As spring progressed they harvested onions, bitterroot, camas, and other perennials. They gathered at traditional fishing sites to share seasonal steelhead and salmon runs. In late summer and fall they ranged into the mountains to follow ripening berries and migrating game.

"The women dug up this plant [bitterroot] with pointed dogwood digging sticks, called pee-cha....The thin, twisted roots are dug when the plant is in bud ... peeled immediately ... dried on racks. When fully dried, they keep a long time."

—Mourning Dove

Colville woman digging bitterroot, c. 1955

"**A very beautiful yellow lichen over the dead brushwood affords a very durable yellow color, and is used by the natives in dyeing.**"

—DAVID DOUGLAS

YAKAMA WOMAN TANNING SKIN

Adeline Fredine, historian with the Colville Confederated tribes, describes how councils would send experienced diggers and pickers out to assess the conditions of various gathering grounds. These scouts would then report to an elder responsible for determining where and when the bands would harvest. By carefully managing these plant resources, they were able to take advantage of the natural fluctuations in yields. David Thompson was the recipient of the tribes' seasonal bounty during his 1811 canoe trip around the Big Bend, when he and his crew of nine depended on food obtained by gift and trade from local bands. Near the mouth of the Okanogan River, the explorer was greeted by a "Chief & about 60 Men and their Women & Children, who made us a Present of 5 Horses, 5 good roasted Salmon, about a bushel of Arrow Wood Berries & about 2 Bushels of Bitter, white & etc. roots—some of them I had never seen before."

After the explorations of Lewis and Clark and David Thompson, British and American fur companies used the Columbia River as a broad highway, establishing strategic trading posts and developing the native people's ancient trails into commercial trade routes. The nature of their business kept the furmen close to the rivers for the most part, whereas the wide open spaces of the interior soon attracted entrepreneurs of a different sort. Many early visitors noted the fitness of tribal horses (Meriwether Lewis pronounced them "as fat as seals"), and several newcomers recognized the value of the native bunchgrass as forage for livestock. By the 1850s, an estimated two hundred thousand cattle were grazing in the Columbia Basin.

When gold strikes in British Columbia during that same decade brought a rush of prospectors and a market for beef to the region, cattlemen were quick to expand their operations. After the first mining frenzy subsided, the demand for beef plunged, and

Cowboy herding cattle in Grand Coulee, c. 1916

many ranchers brought in sheep to take advantage of a market for wool. By 1880 large herds of wild horses also were running loose in the Big Bend. Together these animals took a heavy toll on the native vegetation; written accounts of the time describe areas of overgrazed bunchgrass all around the Basin.

As ranchers and homesteaders replaced the fur economy in eastern Washington, pressure on both native people and the land increased. The Plateau bands did not give up their homes easily, and for thirty years skirmishes alternated with negotiations. For some time, it appeared that the tribes would end up with a sizeable portion of their

By the 1850s, an estimated two hundred thousand cattle were grazing in the Columbia Basin.

As different tribes were shoehorned together on reservations around the periphery of the Basin, their traditional lifeways began to disappear, in the words of Chief Moses, "like bunchgrass before the plow."

BAREFOOT SCHOLARS AT NEW SCHOOLHOUSE IN BURBANK, C. 1909

homeland. In 1878, Chief Moses, a Columbia Salish acting as spokesman for several Basin bands, proposed a reservation that included the river's entire Big Bend, from the lower Spokane to the mouth of the Yakima. This request, acknowledged by General Oliver Otis Howard as reasonable, was vehemently opposed by Yakima Valley settlers and quickly denied by the War Department. As different tribes were shoehorned together on reservations around the periphery of the Basin, their traditional lifeways began to disappear, in the words of Chief Moses, "like bunchgrass before the plow."

Many more plows were soon to come. While some people called this area the "Great Columbia Desert," others saw vast potential. In a surveying report delivered to

Two men working on irrigation sluices, eastern Washington, early 1900s

People dug miles of ditches to water orchards in the Yakima, Klickitat, and Wenatchee Valleys in the foothills of the Cascades.

Congress in 1882, U.S. Army Lieutenant Thomas Symons extolled the attributes of the arid country: "This section has never seemed to enter into the minds of people except as a broken and almost desert land, but I speak from knowledge acquired by traveling over nearly the whole of it, and I shall not hesitate to characterize it as a very fine agricultural and grazing section. . . . The bunch-grass country is the best and finest country on earth, bunch-grass cattle are the sweetest, fleetest, and strongest in the world, and a bunch-grass man is the most superb being in the universe."

The discovery that the rich loess soil of eastern Washington would produce bumper crops drew more people to the region than it had seen in all its thousands of years of

"The wind whistled around the lonely homestead houses, and the dust gathered unmolested in the vacated houses where no woman's hand now kept it wiped away after each dust storm."

—HOMESTEADER LAURA TICE LAGE

human occupation. The era of homesteaders began around the edges of the Columbia Basin, in areas where greater rainfall and deeper soils made farming most feasible. People radiated out from the missionary gardens and cattle pens of Walla Walla; they dug miles of ditches to water orchards in the Yakima, Klickitat, and Wenatchee Valleys in the foothills of the Cascades. Wheatgrowers poured into the grassy hills of the Palouse in the late 1870s. Railroad companies soon began to lay tracks across the harsher central Basin, and their freight-seeking publicists touted the heart of the shrub-steppe as needing only irrigation to blossom into fertile farmland.

By the turn of the century, a grid of small farms and grange halls spotted even the driest reaches of the region. Some of the homesteaders cleared their patches of land and made out for a generation or two, but for those who did not have a reliable source of water, the situation was always tenuous. Without the anchoring roots of the native vegetation, their fields were vulnerable to erosion; spring rains washed silt into streams and creeks, and wind picked up the light soil from plowed fields. Fierce dust storms whirled across the land with increasing frequency, sometimes blowing seeded wheat right out of the ground. Invasive weeds like cheatgrass, brought in during the late 1800s with imported seed, cut into grain yields. In the late teens and through the twenties, severe drought withered crop after crop, and millions of tons of topsoil blew away. Banks foreclosed on homesteads, and many settlers left their land. The exodus continued during the Great Depression of the 1930s.

The hardy souls who hung on learned to calculate their every move within the permutations of climate and topography specific to their piece of ground. Dryland wheat farmers who ply the land today—some of them third- and fourth-generation

LONG-ABANDONED FARMHOUSE

A catalog of the rocks and rivers, sagebrush and salmon—all the pieces that collectively define what we recognize as the Columbia Basin shrub-steppe—provides only a static snapshot of the place. What breathes life and triggers the pulse of the place is a suite of powerful natural forces that have woven together the fabric of this landscape.

Seasonal floods in the Columbia River and its tributaries built bars and scoured new channels. Returning floods of spawning salmon brought an endless stream of food and nutrients from the ocean back into the interior. Lightning ignited fires that raced across the steppe, generating a remarkable patchwork of burned and unburned habitats. Winter snows and spring rains created millions of small, temporary pools in depressions across the Channeled Scablands. For a brief few weeks, these pools teemed with a myriad of vernal flowers and strange animals that could quickly reproduce before shrinking back before the onslaught of summer heat and drought.

In the face of these forces, most human efforts for thousands of years have been notably modest. Occasional fires set by Native Americans no doubt blackened the shrub-steppe. Acres of camas beds were regularly dug and fired as well.

GRAND COULEE DAM

Thousands upon thousands of salmon were harvested. But only in the last century have humans had the means, through machinery and sheer numbers, not only to redirect and manipulate these forces, but in many instances to become a primary force themselves.

Much as the Missoula floods once redrew the face of eastern Washington, through its dams on the Columbia the federal Bureau of Reclamation turned water into a powerful currency of change. A flood of hydroelectric power and irrigation water is again sweeping aside the native vegetation, but this time replacing the shrub-steppe with fruit orchards, vegetable fields, golf courses, and cities. The loess soil of the Palouse is once again on the move, but instead of by wind, more often now it is by water, as the plow-loosened soils are carried away by erosion.

The shrub-steppe is strangely linked to these forces, sustained in extremes of heat and cold, but maintained in a delicate balance between excesses and deficits of fire and water. Too much or too often, too little or too infrequent of either will spell its demise. Despite a remarkable resilience, shrub-steppe only survives today where the human hand rests most lightly on the land.

—*Peter W. Dunwiddie*

"The majestic coulee tells a heroic tale of vanished power and glory far transcending that of Niagara . . . It is now proposed to re-establish the ancient waterway through the great trench of the Grand Coulee. A mere trickle it will be, compared with the river of glacial times."

—J Harlen Bretz

descendants of those first settlers—can recite the exact average rainfall of each producing field, and estimate the results, in bushels, of the slight fluctuations that each year brings. Rancher Jack Linville of Moses Coulee recalled a February drive to the Tri-Cities that points up the variations in effective moisture and growing days across the Basin.

"The snow was six inches deep when I pulled out of my place," Linville said. "It was still falling along the river below Rock Island Dam, but by the time I got to Moses Lake it had started to rain. When I turned off south the sun was trying to peek out, and as I came down on the Columbia again above Pasco everyone had their center pivots sprinkling just as hard as they could go, on crops that were already well along. You think where you live and who controls the water out here doesn't make a difference?"

Water, in fact, has always been the key to tracking people across the shrub-steppe. Soon after the defeated homesteaders departed for a lack of it, New Deal irrigation projects attracted a second wave of immigrants. By the late 1930s, the construction of Grand Coulee Dam alone employed eight thousand men, and the promise of plentiful water refueled the dreams of many farmers. World War II changed the focus of the dam project from irrigation to hydroelectric power, and it was not until 1951 that the massive web of irrigation canals known as the Columbia Basin Project was opened, making good on the mournful refrain of the first homesteaders: "If we just had water, this soil would grow anything." Soon agribusiness consolidated many small farms into consortiums specializing in potatoes, sugar beets, and alfalfa hay; their huge warehouses and storage barns are among the most visible edifices on the steppe today.

World War II also brought the grand scale of the Hanford Nuclear Reservation to the White Bluffs section of the Columbia. As scattered farms, ranches, and small towns

Inspecting wheat in the Palouse Hills

"Owing to the pervasiveness of white man's influences, very few fragments of primeval vegetation are left upon which to base a vegetative study."

—Botanist R. Daubenmire, 1970

were displaced and literally removed from the map, over fifty thousand workers crowded into the new urban center of the Tri-Cities. The security concerns that surrounded the massive reactors had the side effect of creating a large preserve of shrub-steppe, parts of which have been left largely undisturbed for over fifty years.

Today, such places beyond the reach of human development are increasingly hard to find. From paved roads where it is possible to drive for miles without sighting a single clump of sage to sprawling towns that occupy historic encampments along the rivers, people have literally changed the face of the Great Columbia Plain. Laura Tice Lage, who homesteaded near Othello with her parents in 1906, found that she could hardly recognize her childhood home when she returned after an absence of many years. An irrigation canal ran through her former backyard, and everything was green. "Only by looking at the familiar outline of Saddle Mountain lying to the southwest . . . or still farther to the west where the Cascade Mountains and Mount Rainier are etched against the warm colors of sunset, can one be sure this is the same place. Where once sage-bush, bunchgrass, coyotes, jackrabbits, and rattlesnakes had undisputed possession, now stands a modern city."

Although overwhelming changes have been wrought by humans during the last century, reminders of previous waves of settlement remain visible on the shrub-steppe. Gray homesteads slowly crumble in the shadow of walnut and locust trees. Each spring, the charcoal signatures of camas ovens erode from gravel banks alongside swollen rivers. Stone cairns decorated with lichens stand on isolated hilltops. Kiln-dry winds still rattle down off Saddle Mountain, carrying a fresh whiff of sage and the swish of ancient footsteps through the bunchgrass.

CENTER PIVOT IRRIGATION ON ALFALFA AND GRAIN FIELDS

shaded by long grass

PLANT COMMUNITIES

During the spring of 1866, a Massachusetts plant lover named Caroline C. Leighton traveled inland from Portland all the way to Kettle Falls. On her way around the Big Bend at the very peak of wildflower season, she delighted in the subdued colors of "pale salmon and peach-blossom," touched the "delicate fur" of pubescent leaves built to conserve moisture, and passed "great smoking mounds" where native women and children roasted freshly dug roots. As if visiting a marvelous alien planet, Leighton reveled in the strangeness of it all, then sighed with relief at the appearance of the first jagged lines of ponderosa pine near Spokane: "It was like getting to an inhabited country,

"This plain as usual is covered with aromatic shrubs hurbatious plants and a short grass. Many of those plants produce those esculent roots which form a principal part of the subsistence of the natives."

—Capt. Meriwether Lewis, 1806

◀◀ Beezley Hills ◀ Basalt daisy

to reach the trees again: they were almost like human beings, after what we had seen."

One thing Leighton had seen was a lot of sagebrush, and the people she met along the way had little good to say about it. "In this part of the country," she recalled, "sagebrush is a synonym for any thing that is worthless." Ranchers disliked sage because it wasn't good forage for cattle; farmers cursed its deep roots that had to be pulled out before they could plant crops. Yet the native peoples of the Basin found plenty of value in the ubiquitous shrub. They boiled its pungent leaves into a strong tea for colds and sore throats, and brewed a dark suffusion from its seedheads and branch tips for more serious ailments. They braided the loose bark of mature bushes into cordage used for quiver cases, saddle blankets, clothing, and sandals. In their stories, the mythical Coyote often depended on sage, as when he cajoled Raccoon's seven daughters to pile it up for firewood, or when he carried a clump inside Swallowing Monster's belly to burn up the beast from the inside.

Ecologically speaking, sagebrush also has great value. Its tolerance for arid soils, irregular rainfall, and wide temperature fluctuations make it one of the linchpins of shrub-steppe plant communities. Its evergreen branches provide shade for smaller plants, nesting sites for many shrub-steppe birds, and food for grouse, rabbits, and elk. Among several native species of the genus *Artemisia,* the one called big or tall sagebrush covers the most ground, adapting to a wide variety of conditions. In deep soils that receive fair rainfall, big sage stands up as tall as a full-grown man; across dry rocky flats, it straggles along like a crawling baby.

Underground, big sagebrush produces a two-phased root system, sending out a shallow network of small tendrils that can quickly absorb surface moisture, while

Native shrubs along the Columbia River

"These aliens show no evidence of yielding to a reinvasion of the natives once the grazing stops."

—R. Daubenmire

coarser, more robust taproots reach toward deeper reservoirs. This makes it possible for sage to continually draw water from the subsoil and keep on growing during the hot summer months instead of slowing its activity like many other dryland plants. Above ground, big sagebrush produces two different kinds of leaves. The most visible are three-lobed and covered with gray hairs that catch dew and reflect sunlight. These hang on all year, whereas the buds of smaller, softer leaves appear on the branch tips in early winter to take advantage of seasonal moisture, then drop away with summer's drought.

Cousins of big sagebrush appear according to small changes in habitat. Stands of stiff sagebrush, stumpy and serpent-green, grow on harsh frost boils and shattered rocks, while three-tipped sage favors moister soils on northern exposures. Other shrubs that look a lot like *Artemisias,* and mimic many of their adaptations, occur both in pure stands and spotted in with sage: antelope bitterbrush on arid benches; hopsage across a variety of habitats; saltbrush on alkaline flats; winterfat and rabbitbrush depending on changes in elevation.

All of these shrubs tend to grow on wide centers with open space between them. In those spaces sprout spiky tufts of native bunchgrasses, the other key plant family of the shrub-steppe. Bunchgrasses thrive in steppe conditions because their branching roots absorb and conserve water with great efficiency, and their tissues tolerate severe drought with minimal damage. They go dormant during hot summer months, concentrating their growth and flowering in seasons of more reliable moisture. Their reproductive systems depend on elongated stamens and sticky stigmas, and their airborne pollen suits a place with steady wind. The extensive root systems of grasses such as bluebunch wheatgrass, Indian rice grass, Sandberg's bluegrass, needle-and-thread,

hitchhikers and transplants

RUSSIAN THISTLE

Settlers of Washington in the early 1800s brought with them both figurative and literal seeds of change. Contained in their wagons were an assortment of domesticated plants and animals from back home. Unintentionally, they also carried such things as knapweeds and Norway rats—organisms well-suited to flourish in pastures and plowed fields, under porches and in livestock pens.

No one knows exactly how many plants, animals, fungi, bacteria, and other organisms have been introduced into eastern Washington in this manner. The vast majority has stayed closely tied to their human companions, remaining in the flower boxes, pens, and dooryards of their hosts. However some have spread explosively in the wild, at times imparting vast changes to the native ecosystems in which they have successfully inserted themselves. Today, the most abundant plant in much of the shrub-steppe is cheatgrass, an imported grass from Eurasia. The soft green of new cheatgrass seedlings lends a deceptively refreshing wash to entire hillsides each spring. In June, after seizing the shrub-steppe's scarce water supplies, the grass withers and dies—leaving inedible, parched stubble where lush tussocks of native bunchgrasses once flourished.

The list of troublesome, invasive plants and animals grows every year. As non-indigenous species flourish, they impart a homogeneity to much of the formerly diverse shrub-steppe. Arresting and reversing this loss is one of the primary concerns of both private landowners and public agencies concerned with saving Washington's resource lands.

—Peter W. Dunwiddie

"We saw a great many plants of the lupine family, in every variety of shade, from crimson, blue and purple, to white."

—CAROLINE C. LEIGHTON

SULPHUR LUPINE AND LARKSPUR

and prairie junegrass anchor the fine loess soils, and their thick basal tufts shelter small creatures of every description. Birds, mammals, and insects all rely on bunchgrasses for a constant supply of nutritious food that can be browsed on the spot or stored for years at a time.

Walking through a mature stand of sage and bunchgrass is like traversing an old-growth forest, where the trees and understory keep their their own balanced proportions. But while the forest is stretched vertically by tree height, the shrub-steppe is stretched horizontally by open space, and its wide surface is layered with a variety of plants that descend from a modest canopy of shrubs into the rubble of broken stones and powdery soil. The long desert spring, often stretching from February into June, reveals those layers in bursts of soft color. Showy penstemons thrust their trumpets almost as tall as the sagebrush itself, and lupines spread out candelabra in shades from white to purple. An assortment of balsamroots splash whole hillsides with rich yellow sunflowers; lower down, buckwheats lurk in camouflage tones of pink and ochre. Indian paintbrushes throw brilliant flames into the air, while beneath the ground their roots parasitize sagebrush roots for water and nutrients. Lomatiums, low-growing and unremarkable on the surface, store subtle and nutritious tastes in their roots; giant puffballs crack into geodesic patterns that smell faintly of cumin. Below the puffballs, a crust of smaller fungi, mosses, lichens, and algae carpets the earth. These cryptogamic plants—ones that reproduce by spores rather than seeds—form a vibrant skin that provides the genesis for vigorous growth above their heads.

"Just watch this," said Pam Camp, a Bureau of Land Management botanist walking the middle reaches of the Horse Heaven Hills on a hot summer afternoon. She bent

Arrow-leafed balsamroot with lupine and biscuitroot

"The *s/waz-u?* [young shoots of arrow-leaf balsamroot] they gather those just as they come up, and they snip them like this and put it in their basket. And while they're eating fish, they take that and they chew it. And your mouth gets blue."

—Thompson elder Annie York

to Nancy Turner, 1973

"The Men in passing the Ravines broke the Scurf of the Soil; the Dust & Sand rushed down as free to the look as water, pouring down for a considerable time & raising a dusty smoke not to be seen thro'."

—DAVID THOMPSON

CRYPTOGAMIC CRUST ON SOIL, ROCK ISLAND GRADE

toward the dessicated crust and aimed a stream of water from her drinking bottle at what looked like a small fist of clay. Within minutes, the dusty grays of August freshened as the dormant moss began to swell with the surge of liquid, unfolding tiny leaves to flash green. "It's these mosses that start the process," Camp said. "Once they have established themselves on the territory, lichens begin to creep in." She was on her knees now, fingering a pewter-colored saucer of lichen wedged between some clumps of moss. "I always remember this one—*Diploschistes*—because it looks like dried cow chips. And that black knobby one there, it fixes nitrogen—that's what helps condition the soil for larger annuals and perennials."

Camp ran her hand across the ground as she described how the many components of the crust work together to effectively seal off the soil from erosion by wind and water. This composite skin is far from impermeable, however: seeds find nurture in cracks and loose dirt created by frost boils, or in tiny holes bored by solitary bees and wasps, or in burrows clawed by rodents. Right beside my foot, the entrance hole of an ant hill was surrounded by the amber hulls of bunchgrass seeds. Although Pam Camp has spent a couple of decades studying the wildflowers of the shrub-steppe, the cryptogamic crust has opened a fresh realm for exploration. "I'd counted, oh, 125 or 150 different kinds of flowering plants from this ridge," she explained. "Then this student from Oregon State came out here and catalogued 80 different species of lichens and mosses in the soil alone. I've got to think there's a lot more to find out about this place."

Multilayered communities of sage and bunchgrass flow across the steppe, dovetailing with other habitats of remarkable diversity. Jumbled talus slopes are fringed with wild roses, serviceberries, currants, and sweet-smelling syringa, flowering shrubs that take advantage of the seeps that collect at the bases of outcrops and cliffs. Moist depressions harbor small sedges and rushes, willow thickets and fairy rings of wild iris; in early May, camas blooms still turn some of these meadows sky blue. At more lush oases, yellow monkey-flowers dip their heads in the water beneath healthy cattails and groves of aspen. Lithosols are areas of fractured rock with almost no soil at all, but they show off some of the most pleasing displays of spring wildflowers, including a host of buckwheats, beautiful goldenweeds, cushiony phlox, and creeping pink bitterroot, the rock rose. Flat saline playas, glistening white with alkali, sprout with succulent greasewood, saltgrass, and Great Basin rye. Active sand dunes sweep across tracts of

"We passed two great smoking mounds, and, on alighting to investigate, found that we were in the midst of a kamas-field, where a great many Indian women and children were busy digging the root, and roasting it in the earth."

—CAROLINE C. LEIGHTON

CAMAS

White Bluffs bladderpod was described as a species new to science in 1995. Its scientific name, *Lesquerella tuplashensis,* incorporates the Sahaptian name for the White Bluffs, the only place this plant is known to occur.

WHITE BLUFFS BLADDERPOD

countryside, decorated with the red flags of sand dock and the tropical orange blossoms of globe mallow. Tiny earth stars sit right on top of the sand, waiting like ivory smoking pipes to puff out their spores, some of which might land on a small scattering of juniper trees.

Together, these various plant communities strike a delicate balance on the open range; like so many aspects of the shrub-steppe, they tread a fine line between toughness and vulnerability. Grazing practices illustrate their sensitivity with striking clarity. R. Daubenmire, a plant ecologist who began studying eastern Washington's vegetation in the 1940s, postulated that because most large grazing animals vanished from this region in the millenia following the last Ice Age, current plant communities developed in the absence of large herbivores. He documented that it takes years for a tuft of bunchgrass that has been eaten during the peak of its growing season to recover. The sharp hooves of ungulates also take their toll on the crust, and the trails that they cut across the steppe leave the soil open to erosion by wind and water and provide toeholds for invasive weeds. Cheatgrass, for example, has taken over so many acres of bunchgrass that many people now consider it a native plant.

The same forces of change that brought livestock to the shrub-steppe also altered the dynamics of fire. Evidence from prehistoric times indicates that fires here tended to be infrequent and slow-burning because of the lack of continuous fuel. There is no evidence that the tribes managed the sagebrush country by burning as they did on forested lands to the north. Cattle ranchers, however, soon discovered that sagebrush did not resprout from its roots after a fire, while the bunchgrasses not only quickly shot up new tufts but also filled in the blank spots left by roasted sage. When Caroline

Leighton passed through the Basin in 1866, stockmen explained to her how they had used fire to transform the land, so that "many acres formerly covered with sage-brush were now all bunchgrass." All too soon, however, invasive weeds began to encroach on the native bunchgrass and changed the entire system. Some of them, like cheatgrass, were annuals that created continuous carpets of highly flammable fuel as they died back in summer. Fires ignited more readily and burned faster over larger areas, resulting in the destruction of even more sage and fragile cryptogamic crust.

Over the long term, the removal of the sagebrush proved detrimental to the ecosystem. Sage and related shrubs not only provide valuable shade for the grasses, but their deep roots extract important minerals from further underground than grass roots can reach. The birds that nest in the native shrubs control populations of many insects that can damage grasses if left unchecked. Damage to the crust had far-reaching consequences as well. The more ecologists learn about plant communities like the shrub-steppe, the better they understand the interdependence among the different components within the system. From the most nondescript moss in the open crust to the biggest cottonwood by a spring, each species assumes a crucial role within the system, and diversity becomes synonymous with strength. Each of these elements laces itself into the intricate design of the floral tapestry across the landscape. The slightest tug on any thread sends tremors deep within the weave.

Each of these elements laces itself into the intricate design of the floral tapestry across the landscape. The slightest tug on any thread sends tremors deep within the weave.

a buzz in the air

INSECTS AND ARACHNIDS

Insects supply the throbbing pulse to sagebrush and basalt; alongside the eight-legged arachnids, they suffuse every layer of the shrub-steppe with teeming life. Solitary bees and wasps, glistening like winged jewelry wire, pop in and out of holes that riddle the cryptogamic crust. They helicopter from one flower bloom to another, joining a host of other tongues in the search for nectar. Ticks and bugs dangle off leaf ends with infinite patience, waiting to hitch a ride. Darkling beetles bumble between geometric mounds that house whole planets of ants; the interiors of their colonies bulge with preserved seeds and incubating eggs. Jerusalem crickets probe underground tunnels for prey

Dragonflies snag midges above ponds and puddles, and every evening spiders weave a new veil of silk over the entire scene.

DAMSELFLY

◄◄ PARACOTALPA BEETLE ON FERNLEAF
DESERT PARSLEY
◄ COMMON BLUE BUTTERFLIES

while scorpions wave their tail stingers in dark recesses. Dragonflies snag midges above ponds and puddles, and every evening spiders weave a new veil of silk over the entire scene. Day and night, the backbeat of insect stridulation rises in brief crescendos that dominate the scene for a relative moment, then fade into a quiet harmony.

Over time, untold thousands of these creatures have developed along with plant partners in a stuttered dance of coevolution. With myriad colors, shapes, and patterns, flowers guide insects to the dusty pollen grains that ensure their propagation; in return, pollen and nectar supply the carbohydrates and proteins in many insect diets. Some insects are generalists, feeding on a variety of flowers or greenery. Others are quite specific in their tastes. Indra swallowtails, for example, lay their eggs on desert parsleys, while the larvae of Behr's hairstreak butterflies feed only on antelope bitterbrush. A bright orange monarch careening through a coulee might be either a migrant on its way to faraway places or a Columbia Basin breeder in search of a patch of milkweed.

The life cycle of the much less common sagebrush sheepmoth depends on expanses of big sagebrush. These black and white cousins of the luna moth, born within a family famous for nocturnal courtship, perform their mating flights during a few hot cloudless days in August. The scent of a perched female can attract males from up to seven miles away, inducing mating flights of legendary proportions. The females lay their eggs on sage plants, and the larvae feed on the leaves when they hatch. The presence of these aerobatic moths supplies one sure measure of an extensive, healthy sage community.

In contrast to the large territory required by sheepmoths, some insects thrive in astonishingly restricted niches over great spans of time. The scrubby hills above the

MONARCH BUTTERFLY

A bright orange monarch careening through a coulee might be either a migrant on its way to faraway places or a Columbia Basin breeder in search of a patch of milkweed.

Yakima River provide homes for a certain kind of leafhopper that is found nowhere else. This small population belongs to a much larger group of busy bugs that hop about, sucking the juice from green stems and shooting drops of sweet honeydew with remarkable rapidity. While other leafhoppers can travel great distances, these Yakima hoppers have maintained simple and sedentary ways, never moving off the hummock occupied by their ancestors while watercourses have cut down through the hills and separated their habitat into distinct units. Islands of these primeval leafhoppers trace changes in the path of the Columbia River over millions of years.

Entomologist Richard Zack of Washington State University has a fondness for the way that sheepmoths and leafhoppers reveal the inner workings of a system. His own specialty centers around the shore flies that inhabit the edges of alkaline ponds, and during the course of his work netting and identifying fly specimens, Zack finds himself

❦ **"Most children seem to have a natural bent toward biology; and watching those locusts from the time they left the ground in the pupa stage through to flight, was a never ending source of interest to us. We would find the cocoons, split down the back, clinging to a branch of sagebrush while the insect struggled to draw its body and legs out of the thin shell that encased it."**

—LAURA TICE LAGE

surrounded not only by untold numbers of insects, but by a host of other life forms as well. "Why do you think so many different birds come to these ponds?" he asks. "It's not just the water. They're here to gorge on the insects that live in and around the water."

Among the insects that attract the birds are grasshoppers, which abound around the edges of such lakes. But then grasshoppers abound in any open situation where there is greenery to eat—they are often the most noticeable of all the shrub-steppe insects. Their successive generations, or instars, multiply through the summer depending on some correlation of rainfall, humidity, plant growth, heat, and wind. By the end of a warm season, a few steps across a sage flat can raise a plague of rasping locusts, and predators from every link in the food chain—from kestrels and coyotes to rattlesnakes and bobcats—feast on this loud and abundant resource.

In the rocky canyons of the Klickitat River, the plant and animal kingdoms are linked in quieter ways by a minute, metallic-blue wasp that deposits its single egg on the emerging leaf of a Garry oak tree. Secretions of the hatched larva coax the leaf cells to overgrow around it, enclosing the soft grub in a perfectly round, weightless, nutrient-rich gall. In turn, this oak apple gall may be invaded by a parasitic fly, or pecked open by a downy woodpecker, or snatched free and stored by a foraging gray squirrel. If the gall should happen to remain unbroken through the summer, an adult wasp bores an escape hole, then burrows down toward the roots of the tree to overwinter and begin the next phase of its improbable life.

Richard Zack finds it easy to be amazed by the permutations of grasshoppers and gall wasps, but more difficult to name them by species. In fact, no one person knows the insects of the sagebrush country very well, and until their names and their roles in

GRASSHOPPER ON GAILLARDIA

Among the insects that attract the birds are grasshoppers . . . but then grasshoppers abound in any open situation where there is greenery to eat—they are often the most noticeable of all the shrub-steppe insects.

Tiger beetles are small, beautifully marked hunters with slender legs and lightning moves. The adults aggressively pursue other insects on the wing.

WESTERN TIGER BEETLE

the ecosystem are more thoroughly understood, many of the subtleties of the whole community will remain mysterious.

"In the long run," offers Zack, "there's a good chance that the insects may be able to tell us more about how the shrub-steppe works than any other single constituent, even plants. But now when I go out after dung beetles, I have to catch twenty-five specimens that look the same, line them up on a board, then put each one under a microscope and start counting hairs before I have any idea whether they are the same or different species. We don't have a feel yet for their diversity; our knowledge is just not there."

Since 1993 Zack, with support from The Nature Conservancy and the Department of Energy, has concentrated on gathering and cataloging insects around the Hanford Nuclear Site. Employing a variety of trapping and preserving techniques, he ships many specimens to colleagues for identification. The inventory's 1997 annual report listed more than sixty-eight authorities from all over the world who have lent a hand to this benchmark local classification. The numbers on the chart—220 species of beetles, 311 flies, 362 bees and ants, 318 moths—can startle the mind, and Zack does not appear to be joking when he estimates that he will be sending in reports and corrections on his collection for the next few decades.

"I see these insects as a window to the past, and not just for the fifty years that they have been left alone at Hanford. Many of their populations go back to a time when sagebrush and bunchgrass communities extended unbroken north up the Okanogan and south into Nevada and beyond. It's what once covered the West, and we never knew anything about what insects were living here, or what they were doing. In some places, there's still a chance to find out."

The shadow of the Cold War reached far into the remote regions of central Washington with the development of the Hanford Nuclear Reservation in 1943. The secrecy that enveloped this project, initially for the manufacture of plutonium for the first atomic bomb, effectively sealed a 560-square mile area from the public for over fifty years.

Activities associated with building and operating a series of nuclear reactors at the Reservation generated copious quantities of hazardous radioactive waste. However, by creating this bio-hazard, they also removed the primary factors—agricultural tilling, grazing livestock, urban development—that elsewhere in the Columbia Basin accounted for the greatest losses of shrub-steppe habitat. Easily distinguishable in satellite photos from the surrounding farmland, the site is now Washington's largest remnant of this once-ubiquitous habitat.

It's hard to comprehend how nuclear weapons production could lead to the preservation of native habitats, or to accept that toxic waste could share a site with rare species that occur nowhere else in the world. Nonetheless, such is the case at Hanford Nuclear Reservation. Recent surveys of Hanford by The Nature Conservancy have revealed dozens of native Northwest species previously undescribed by scientists, plus an assortment of locales occupied by rare birds, insects, and plants.

Numerous interests are currently competing for use of this unusual site. Conservationists are seeking protection for this unique ecosystem, while others have put forth proposals that would allow various agricultural activities. Will fruit orchards and asparagus fields obliterate the biological treasures that military secrecy and toxic waste have inadvertently kept alive? As of this writing, the future of the Hanford Nuclear Reservation is not clear.

—*Peter W. Dunwiddie*

HANFORD B REACTOR FROM GABLE BUTTE

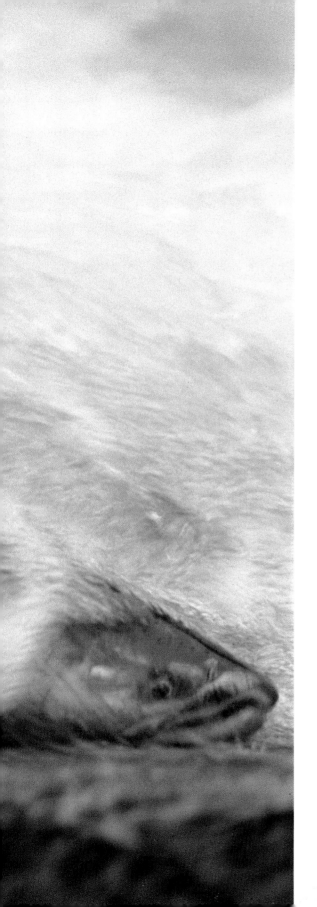

the delicate perceptions of swimmers

FISH

As large grazing mammals progressively diminished across the shrub-steppe in the millenia following the end of the last Ice Age, the hunters of the Plateau adjusted their diet to a resource of comparable abundance in the Columbia drainage. That resource was fish, and the stories of land and water dwellers here have been interwoven ever since.

Numerous native species bred in local waterways, their availability changing with the seasons. For several Basin bands, bony suckers provided succulent white meat to

❧ **"Large log scaffolds extended over the foaming river where men stood to spear or net fish fighting to get up the falls. Some men had set up big baskets woven of red willow with serviceberry rims."**

—MOURNING DOVE

STEELHEAD TROUT

❧ **"Met with another species of trout of a dark color on the back, steelhead salmon trout"**

—MERIWETHER LEWIS

◀◀ CHINOOK SALMON SPAWNING
◀ DIP-NET FISHING ON THE KLICKITAT RIVER
NEAR LYLE

accompany the desert parsley shoots gathered at the beginning of spring. In Rocky Ford Creek, spawning suckers from Moses Lake were diverted by barriers of brush and tules, then scooped by hand from shallow water. Native trout caught below Dry Falls were steamed in small craters filled with heated rocks. In the bare foothills of the Cascades, ravenous red-sided shiners lunged at baitless hooks. Lamprey "eels," who use their sucker mouths to ascend the most fearful rapids, were pried off rocks and draped across poles to dry for winter food. Bottom-dwelling Pacific sturgeon, some of monstrous size, provided a rich source of protein, oil, and scary stories.

Balanced against all these worthy fish, the various species of salmon reigned supreme in terms of sheer numbers and symbolic power. Archaeological digs at The Dalles show that a salmon fishery existed there at least nine thousand years ago. As the explorer David Thompson made his way down the Columbia in July of 1811, he traded for salmon with every band he met. He watched people divert the fish with weirs, capture them in huge woven baskets, and spear them from rocks. He observed the complex rituals that attended the harvest and concluded that "experience has taught them the delicate perceptions of this fish."

The Pacific salmon that we know today evolved from a large freshwater species that lived in inland lakes and streams. During the Ice Ages, some of these fish began riding swollen creeks and rivers downstream into estuaries whose salinity had been reduced by enormous volumes of glacial meltwater. There they developed a tolerance for salt, and eventually ventured far offshore to feast on the abundance of the open sea. The ancestors of our modern species returned to fresh water to breed, developing an anadromous, or two-phased, life cycle.

FISHING IN FRONT OF THE DALLES DAM

GAFFING A CHINOOK

Salmon begin the first phase of life as small fry on the bottoms of rivers and streams. On a schedule that varies from a few weeks to four years, the fry develop into smolt and head downstream toward the ocean. Their saltwater sojourn lasts from one to four years, when a synchronized homing instinct turns the heads of an entire hatch back toward land. Once they reach the mouth of the river they originally descended, the salmon thrash upstream, sometimes for hundreds of miles, until they arrive at the exact place of their birth. They bed their eggs in the gravelly bottoms of their natal creeks and lakes, then die. Captain William Clark, writing from the confluence of the Snake and Columbia Rivers in mid-October of 1805, was overwhelmed by the scene: "The number of dead Salmon on the Shores & floating in the river is incrediable to Say." The carcasses of these salmon provide significant nutrients to the entire food chain of the drainage, from the algae on the bottoms of the streams to the carrion eaters that feast on the rotting flesh.

In times past, the epic journeys of these fish set the calendar for tribes and wildlife alike: native fisheries, bear movements, eagle migrations, and great gatherings of ravens all coalesced around the staggered runs of fish. Historically, five kinds of salmon—chinook (king), sockeye, coho (silver), chum (dog), and steelhead—spawned along the mid-Columbia and its tributaries. Biologists estimate that as many as sixteen million salmon entered the mouth of the big river in the early years of the nineteenth century, but by 1900 the runs were noticeably smaller. Early irrigation projects dried up many small creeks, and runoff from mining and logging operations smothered spawning beds with silt. Cattle and sheep trampled the shrubs and small trees that provided necessary

"The salmon continue to arrive in almost incredible numbers for nearly two months; in fact, there is one continuous body of them, more resembling a flock of birds than anything else in their extrodinary leap up the falls, beginning at sunrise and ceasing at the approach of night."

—Artist Paul Kane, 1847

Salmon smolt swimming against the current

"It is the popular belief that all the Salmon that enter the River die and not one ever returns . . . exhausted by their long Journey, weakened by spawning, they become so emaciated as to have no strength to return and die in great numbers of very weakness."

—David Thompson

Expired salmon in water after spawning

shade along watercourses. The arrival of commercial canneries took a further toll on fish populations as production increased from four thousand cases of canned salmon in 1866 to over six hundred thousand cases in 1883. Beginning early in the twentieth century, the construction of dams created formidable barriers to both ascending adults and descending smolt. The decision not to build fish ladders around Grand Coulee Dam eradicated the fish from the upper river in 1941.

Today, the only natural spawning grounds left on the main trunk of the Columbia are found in a fifty-mile free-flowing stretch of river known as Hanford Reach. Chinook salmon breed along the whole length of the Reach, especially off gravel bars and islands below the White Bluffs. Their spawning beds (called redds) are visible in the river as light-colored areas cleared of gravel by flailing bellies. Each year, smolt from

ROCKY REACH DAM FISH LADDER

"Infinitely greater numbers of salmon could be readily taken here, if it were desired; but, as the chief considerately remarked to me, if he were to take all that came up, there would be none left for the Indians on the upper part of the river; so that they content themselves with supplying their own wants."

—PAUL KANE

BURBOT

Burbot is a cold-water species still caught by anglers in Lake Roosevelt, but they were cut off from the mid-Columbia by the closing of Grand Coulee Dam in 1941.

"Coyote said, 'I am going to get salmon for the people living in the inland country'... He was cunning. He could tell the future, and also he used his magic ways."

—Yakama elder Alva Claparty

to L.V. McWhorter

Chinook salmon

the wild redds and from the Priest Rapids hatchery make their way downstream through the succession of dams. While their numbers fluctuate like those of any salmon run, researchers generally count the chinook in the tens of thousands, and a 1996 Journal of Fisheries survey rated the Hanford Reach fall chinook run as the most productive fishery in the entire Pacific Northwest. Through the early 1990s, fish from this run accounted for a third or more of the combined commercial, sport, and tribal salmon harvested on the lower Columbia. This is a contribution far out of proportion to the size of the Reach in relation to the whole river, and points up the magnitude of troubles upstream and down.

Of all the native fauna of the Columbia Basin, none has been more dramatically affected by modern development than the fish. Introductions of new species and alterations of original habitat have established entirely new ecosystems in most bodies of water. But around Hanford Reach, the list of natives still includes not only spawning chinook and four other migrating salmon species, but also lamprey, sturgeon, Dolly Varden, steelhead, chiselmouth, red-sided shiner, northern squawfish, mountain whitefish, three different kinds of suckers, and five sculpin.

Unfortunately, this cornucopia of fish faces several serious threats. Expansion of agricultural operations on the North Slope could percolate enough water through the soil to undermine the fragile White Bluffs formation, which in turn would alter the configuration of the river bottom at key spawning locations. Just across the river, the Hanford Nuclear Reservation must deal with the containment of daunting amounts of radioactive waste. Sprawling growth from the Tri-Cities area will soon touch the downstream end of the Reach.

In spite of these problems, fish biologist David Geist of the Pacific Northwest National Laboratory believes that as long as the system of bars, currents, and water quality maintains its integrity, both the salmon and their lesser-known kin have a chance to continue their success. A proposal to designate Hanford Reach as a Wild and Scenic River, stalled for several years within the political labyrinth, would go a long way toward accomplishing these ends. "If the local economy can develop in a way that preserves natural capital in the Hanford Reach ecosystem," writes Geist, "then the Reach's human and natural economies stand a chance of being sustainable. If the Reach is managed for sustainability, species that are doing well will remain a reservoir for restoring other parts of the basin."

"They say that the Snow falls only abt 1 1/2 ft deep & soon goes off, there are plenty of . . . Deer, with small Trout & another small Fish in the Winter, which with the dried Salmon forms their livelihood."

—David Thompson at Priest Rapids

Mountain whitefish

sleeping with lizards

AMPHIBIANS AND REPTILES

At dusk on a May evening, along a trickling creek that drains a pothole lake, the air is filled with a cacophony of off-kilter quacks. Energized by a spring shower, multitudes of Great Basin spadefoot toads have emerged from hibernation and are calling for mates. Surrounded by such noise, it is tempting to try and catch one, to touch the black horny blades of its rear digging feet, but their ventriloquistic calls bounce from too many directions, and fade too quickly. The wily toads allow an intruder to approach only so close before sinking beneath the water like baby submarines.

🌿 **"I woke up this morning and found something cold and clammy against my leg. Threw the blanket off and found a reptile of the lizard species who had been my companion during the night."**

—PAUL KANE

NIGHT SNAKE

The spadefoot breeding cycle flashes by quickly, taking advantage of seasonal pools and streams. As males collect in the water, their chorus attracts females, who lay gelatinous clusters of eggs that number in the hundreds. Attached to vegetation in shallow water, these eggs undergo the most rapid rates of development measured anywhere on the continent: embryos hatch within a few days, and within two or three weeks they have grown into algae-eating tadpoles. About the time that spring watercourses begin to dry up, the tadpoles' gills and fins metamorphose into lungs and legs, as the spadefoots recapitulate their ancient evolutionary leap from water onto land. The young toads hop away to begin terrestrial life chasing after crawling insects and fleeing from all manner of predators. When the drought and heat of summer arrive in earnest, the toads use their peerless spades to dig with remarkable speed down into the sweet cool depths of dark earth. There they sometimes remain buried in suspended animation for eight months or more, until the next spring downpour calls them back to the surface.

Although torrid, rainless summers would seem to work against moisture-loving amphibians, several different frogs, toads, and salamanders use strategies similar to those of the spadefoot to survive and even thrive throughout the Basin. Beneath a small waterfall, within sight of the toads' humble creek, live populations of the two salamanders who inhabit the shrub-steppe. Along the edges of the pool move the tiger salamander, beautifully marbled in shades of green and black, and the long-toed salamander, its star-speckled body marked by an ochre stripe down the middle of its back. Both of these "water dogs" breed in still, shallow water. Tadpole-sized larvae emerge after a few weeks and hang near the surface, their bodies the color of greenish duckweed, with threadbare legs and tiny golden eyes. Three long, feathered gills trail behind

◄◄ SAGEBRUSH LIZARD
◄ GREAT BASIN SPADEFOOT TOAD
EMERGING FROM THE SOIL

BLOTCHED TIGER SALAMANDER LARVA

gaping mouths that quickly gobble worms, swimming insects, tadpoles, and others of their own kind. At the end of one or more summers in the water, most larvae metamorphose into adult salamanders and migrate to shore, where they slip into underground tunnels and burrows.

For years the main students of such amphibian life histories were bait-seeking fishermen and curious children chasing polywogs. More recently, however, the desire to see and understand these supple creatures has gained a sense of urgency, because so many of them, in so many places, appear to be fading away. Possible causes for this slump in numbers range from climatic changes through toxic agricultural runoff to the introduction of non-native species. The omniverous common bullfrog, for example, which has been imported from the eastern United States, devours other amphibians at an alarming rate and has seriously damaged populations of both spotted and northern

A western skink disappears into a narrow fissure of cinnamon basalt, leaving its sky-blue tail wriggling in the beak of a magpie.

WESTERN SKINK

leopard frogs. Similarly, certain introduced fish consume the eggs, larvae, and adults of several amphibian species. But the overriding factor, as in so many cases of decline, appears to be loss of habitat—the filling in and paving over of the wetlands that are absolutely essential to amphibians during the rigors of their dryland year.

The most common behavior of all herptiles—the amphibians and reptiles—involves thermoregulation, because these cold-blooded creatures can only operate within a relatively narrow temperature range. While amphibians take comfort in coolness, darkness, and moisture, their reptilian cousins require warmth in order to function. Guided by specialized heat-sensing organs, during the course of each day they gravitate between sun-baked surfaces and shady crevices in order to maintain their optimal temperature.

Reptiles also put their bodies through all sorts of contortions in their constant efforts to stay warm enough, but not too warm. Snakes knot themselves into sinuous coils to conserve heat, or stretch out into rigid sticks to lose it. A western fence lizard will perform strenuous pushups and neck bends, then shuttle the very pigment across its scales, shading its body color from light to dark as the situation demands. In broiling heat, a sagebrush lizard will curl its toes to minimize contact with a hot surface. During winter weather, shrub-steppe reptiles seek deep shelter in order to survive, and several kinds of snakes—including rattlers, gophers, racers, and whipsnakes—have been observed to share rocky natural dens called hibernaculums.

Biologists list a single turtle, five lizards, and nine snakes as native to the Columbia Plateau. These vary from western painted turtles that never venture away from water for very long to blue-tailed skinks who seem comfortable under any condition, wet or dry;

SHORT-HORNED LIZARD

"Examined a Species of Lizzard Called by the French engages, Prairie buffaloe. Above and behind the eyes there are Several Projections of the bone which . . . has the appearance of horns Sprouting out from the head. This part has induced me to distinguish it by the appellation of the Horned Lizard."

—MERIWETHER LEWIS

from night snakes so reclusive that humans rarely glimpse their haunting vertical pupils to side-blotched lizards that continue about their business with a watcher only a few feet away. Of all these species, two in particular seem to epitomize the shrub-steppe.

With its stubby goat horns, long, grasping digits, and mouth downcast into a permanent grimace, the short-horned lizard carries a classic reptilian look. On a hot afternoon in the dunes, this lizard might lie right out in the open, the colors of the shrub-steppe encoded in its scales: lichen oranges and yellows, basalt browns and grays. About the size and shape of a buccaneer's medallion, this tiny dragon can sometimes be seen tracking a line of golden ants, its favored food, along an outcrop of basalt. When threatened, a short-horned can flatten its body until the toothed fringe between its back and belly blends seamlessly into the surface. Like all lizards, it has sensitive ears, sharp eyes, and effective claws; unlike most lizards, it does not lay eggs but gives birth to between three and fifteen live young each summer. Sahaptian people compared

Sahaptian people compared short-horneds to Indian doctors who cured patients by blowing on afflicted body parts. They spoke of this small creature's ability to whip up the winds of bad weather, and accorded it the kind of respect they offered the more imposing rattlesnake.

short-horneds to Indian doctors who cured patients by blowing on afflicted body parts. They spoke of this small creature's ability to whip up the winds of bad weather, and accorded it the kind of respect they offered the more imposing rattlesnake.

Rattlers have always been the best known and most feared of reptiles, and the western rattlesnake is no exception. Accounts of early travelers and homesteaders describe many a nervous walk through the sagebrush, and most settlers made a point of killing every snake they saw. Reasonable caution combined with exaggerated fears to create an image of rattlers as evil vipers lying in wait to strike at passing humans. In reality, rattlesnakes are shy and reclusive creatures. When approached, a single snake usually freezes or retreats, buzzing its tail only if an intruder comes too close. The small number of bites that occur each year usually result from sudden encounters in tight quarters or from people attempting to handle snakes. Greater knowledge of the habits of pit vipers might have changed the opinions of some of those early homesteaders—they are common snakes, and they consume huge numbers of rodents and ground squirrels each year. The superstitions surrounding rattlesnakes often carry over to their nonpoisonous kin as well, and as a result, plenty of harmless gopher snakes and racers have been chopped into tiny pieces by garden hoes. While it is difficult to overcome the myths perpetuated by misinformation, there is no better place to accomplish this—that is, to watch what reptiles really do with their time—than in the open shrub-steppe.

As the sun climbs in the sky on a summer morning, a western painted turtle, its shell rimmed with fiery red, hoists itself out of the water onto a table rock to bask away the day among a flurry of shore flies. A little rubber boa, caught unexpectedly out in the daylight, bundles its body into a shapeless clump and releases a cloud of pungent

STRIPED WHIPSNAKE

musk. A western skink disappears into a narrow fissure of cinnamon basalt, leaving its sky-blue tail wriggling in the beak of a magpie. A desert striped whipsnake, one of the rarest serpents of the shrub-steppe, ascends a clump of saltbush, winding its long, slender, sinuous body through the branches like a beautiful ribbon. The sight of an airborne reptile the length of a man sends waves of alarm and fascination through the mind of an approaching walker, but the whipsnake feels no such drama—on such a hot day, it only wants to expose more of its belly to the breeze. Unravelling itself from its perch, the snake launches its head into space and then gently lowers it onto the crusty ground. The striped body follows in perfectly controlled loops, feeling a path among rocks; with a casual flick of its impossibly long tail, the whipsnake blends into the bunchgrass and disappears.

tracks and dung

MAMMALS

To spend a night in the shrub-steppe is to give yourself over to the world of small mammals. At dusk the face of an empty sand dune, wiped clear by the afternoon wind, suddenly fills with footprints. Dark eyes peer out of basalt cracks, and bats swoop away from secret roosts to dine on ascending swarms of insects. Grasshopper mice howl and yip like tiny wolves, attacking everything from scorpions to other mice. On cliffs and talus slopes, bushy-tailed woodrats drum their hind feet in syncopated patterns that echo through dreams. Among hummocks of bunchgrass, black-tailed jackrabbits chew on fresh sprouts, while finger-sized Merriam shrews sniff out beetles for food. As Great

"The American badger is very abundant in the plain country east of the Cascade mountains, its burrows perforating the ground thickly in many places to the great danger of both horses and riders."

—BIOLOGIST J.G. COOPER, 1854

Basin pocket mice and least chipmunks stuff their cheeks with seeds, sagebrush voles scramble up from their nests of sage bark to nibble at the freshest flower petals in their desert garden. Not even the naturalist David Douglas, who fancied himself quite a rodent enthusiast, was prepared for the intensity of a July night in the Columbia Basin. At a camp beside the river in 1826, he found himself "annoyed by the visit of a herd of rats, which devoured every particle of seed I had collected, eat clean through a bundle of dried plants, and carried off my soap-brush and razor!"

There is method to the mad activities of these small marauders. In lives that span the entire breadth of shrub-steppe habitats, they provide lucid examples of adaptations to arid situations. Many of these small mammals metabolize their water—all of it—from a diet of seeds and plant parts. During the warm months, most of them do their hunting and gathering between dusk and dawn, then spend the heat of the day napping in shady crevices or underground burrows. Pocket gophers use their powerful shoulders and blunt heads to bulldoze ropey mazes, where they wander safely ensconced in darkness. Pocket mice camouflage their entrance holes and control the daytime humidity in their tunnels with plugs of grass clippings. Townsend's ground squirrels and yellow-bellied marmots conserve moisture and escape extreme summer heat by estivating in the cool recesses of their burrows and dens. Since these squirrels also hibernate during the winter, they can remain dormant for seven or eight months of the year. Jackrabbbits, instead of going underground in the cold months, change their diet from summer greenery to bark and twigs and keep hopping. Sagebrush voles, who remain active during all hours of the day and all seasons of the year, bore runways through the ground vegetation to protect themselves from both predators and extreme

◄◄ TWO BADGERS
◄ BLACK-TAILED JACKRABBIT

COYOTE AMONG SAGEBRUSH IN WINTER

temperatures; they have even been known to tunnel through dried cow chips to expand their system.

Two larger rodents that depend on watery habitats also have long-established niches in the shrub-steppe. Muskrat diggings riddle the banks of rivers and lakes wherever the vegetation grows vigorously enough to feed them. Beavers thrive in wetlands ringed by aspens and cottonwoods. Another pair of medium-sized mammals, the raccoon and the

"Coyote would go about hunting; he would shoot and kill all sorts of things."

—YAKAMA ELDER JOE HUNT TO MELVILLE JACOBS, 1929

"At 2:30 pm saw the first sheep—
Michel went after it, but the wind had
started it."

—DAVID THOMPSON, JUST BELOW THE

MOUTH OF THE METHOW RIVER

skunk, are both familiar and omnivorous. Their range often follows the spread of human settlement, and both are increasing in certain areas of the Basin.

Although most of the carnivorous mustelids (members of the weasel family) are normally associated with running water and dense cover, long- and short-tailed weasels, mink, and even river otters have been known to make appearances along coulee watercourses. The badger, a mustelid that does not need any water at all, stands out as a formidable predator of rodents. Wherever the soil is deep enough, loose mounds around elliptical entrance holes testify to the presence of these relentless diggers. A chorus of anxious ground squirrel whistles, accompanied by the sight of dirt being tossed high into the air, signals that a serious hunt is in progress. The long claws and stubby, powerful limbs of the badger excavate at a furious pace, and portend no good for any small rodent cowering at the end of the tunnel.

Compared to the singleminded style of the solitary badger, the social coyote seems almost leisurely in its ways. Its keen eyesight and sense of smell, combined with the cooperative strategies of the pack, allow the song dog to fit perfectly into the broad sweep of the shrub-steppe. Coyotes will eat anything from crickets to carrion, and have been known to cache uneaten food for hungrier days. They den in either soil or rocks, and can lope along at a steady pace for miles on end. They seem to be able to weather hard times: the sight of a skinny coyote, touched with mange and staggering along with its nose to the ground in search of crawling insects, is just as familiar as the image of a well-fed one, chestnut and sleek, contentedly watching the countryside from the shade of a basalt overhang. Having survived poisoning and bounty campaigns aimed at its complete eradication, the coyote still exhibits the same qualities of cunning and

Bighorn rams

While the historic range of elk in the Columbia Basin is poorly understood, tribes like the Yakama have an entire range of words describing their growth and habits, and there is no question that wapiti are presently expanding their range in the shrub-steppe.

resourcefulness that made its mythical counterpart such an irresistible character to generations of Plateau-culture storytellers.

Cats also lurk in the Basin. Bobcats are known to dwell within abandoned reactor buildings at Hanford, where there is no shortage of small birds and rodents to eat. In recent years cougars, perfectly camouflaged in certain light sands and butterscotch basalt, have been sighted around bare mesas such as Gable Mountain. Their traditional prey are the hoofed grazers of the shrub-steppe: strapping mule deer that clatter up and down talus slopes in every coulee, and elk that have been wandering into the Basin from the Cascade foothills and the highlands around Spokane with increasing frequency.

Population trends among many of these native mammals are often hard to interpret, and no group presents more of a puzzle than the various dryland rabbits. One of them, the mountain cottontail, forages along narrow trails, feeding on sage, bunchgrass, and perennial herbs. This is not a particularly fast or elusive rabbit, and when faced with danger it often freezes where it stands or takes a few lippity hops into the cover of a bush or rock crevice. Even so, it escapes from marsh hawks and coyotes often enough to stand as the most successful rabbit over large portions of a range that extends from woodlands in the Southwest, across the Great Basin sagebrush, and well up into the mountains of eastern Washington.

Jackrabbits, on the other hand, tend to be more specific in their habitats. Black-tailed jacks are usually found in lower-elevation sagebrush, while white-tailed favor slightly higher elevations and more mixed vegetation. With their sinewy physiques, broad foraging habits, and wildly athletic leaping abilities, jackrabbits seem perfectly suited to the rugged steppe. After the arrival of homesteaders, the jacks quickly

Elk in winter

"These hares are so numerous that our command of 60 men subsisted on them for nearly a week. The flesh is rather bitter, owing probably to the sage on which it feeds."

—BIOLOGIST GEORGE SUCKLEY, 1854

developed a fondness for potatoes and hay. As bounties reduced the numbers of their coyote predators, the rabbits became such a nuisance that during the early 1900s settlers organized drives that killed many thousands of them. The farmers' efforts were temporarily successful, but after a brief hiatus the jacks moved back in. Yet in more recent decades, sightings of both of these hares have dwindled. Patches of desert phlox, once nibbled by jackrabbits into neat hemispheric clumps, now hang in straggling disarray, and researchers wonder if there are enough reserves of either jackrabbits or habitat to rejuvenate two species whose numbers are clearly on the wane.

Even less common than the jacks is the pygmy rabbit. This smallest of all rabbits is colored a pearly gray right down to its cottony tail, which works as an effective camouflage around shadowy clumps of sage. The life of a pygmy is built around tight parameters that include deep, loose soil for digging and healthy sagebrush mixed with shrubs like rabbitbrush and greasewood; it feeds on sagebrush year-round, and seldom wanders more than thirty yards from its home burrow. Sometimes in the evening, its faint barking sounds can be heard rising from the lip of an entry hole.

Bill Rickard of the Arid Lands Ecology Reserve on Rattlesnake Mountain thinks that the habitat of this diminutive rabbit is comparable to that of another difficult-to-monitor species, the sagebrush vole—both seem to fare best in mixed sagebrush and bunchgrass communities on slightly higher elevations that receive a little more rainfall than the flatlands. Pygmy rabbits and voles appear regularly in suitable habitat in east-central Idaho, but in eastern Washington their numbers are spotty enough to mark them both as species to monitor.

It is certainly possible for animals like this to slip entirely away. Pronghorn antelope—whose bones can be counted by the thousands around prehistoric campsites from the Okanogan to the Snake—bounded across the steppe until around the time of European contact, then disappeared, probably due to some combination of climatic change, vegetative mix, and human predation. Early travelers sighted bighorn sheep on the rim of Grand Coulee, but there are none there now. The coyote, on the contrary, which coexisted with the antelope and sheep for untold millenia, still ranges across the entire steppe, serving as a reminder that it is never easy to predict the reaction of an animal to change. Other mammals that have also felt pressure from human encroachment play a regular and even increasing role among the current fauna of the region.

When Bill Rickard saw a handful of elk at a large spring below Rattlesnake Mountain in 1972, he thought they must be wanderers or wintering animals. "The experts said they needed shade," recalls Rickard. "Then we got some graduate students out here who watched them lie down right out in the sun." A breeding population seemed unlikely; research papers maintained that the nutrients in drygrass vegetation would lead to underweight calves and cows with inadequate milk. Rickard read the reports and watched the animals. In a situation like this, he says, "You document everything as carefully as you can, then pass it on to the next generation."

The first elk calf, born on the Arid Lands Ecology Reserve in 1982, was followed by others that weighed in a little heavier than average calves from the Blue Mountains. Today there are about six hundred elk on the Arid Lands Ecology Reserve, where they browse on sage and nibble on peachtree willow shoots around Rattlesnake Springs. Rickard declines to speculate on the future of the herd, but remains keenly interested in watching what they do.

The coyote, on the contrary, which coexisted with the antelope and sheep for untold millenia, still ranges across the entire steppe, serving as a reminder that it is never easy to predict the reaction of an animal to change.

wild cries of wild fowl

BIRDS

It is impossible to travel through any part of the Columbia Plateau without sensing the flick and soar of many wings. The expansive shape of the land provides a magnetically attractive space in which to fly. Fractured basalt pinnacles serve as peerless lookout roosts, while long low ridges and interconnected coulees offer navigation guides for migrants. Infinite nicks, caves, pothole lakes, evergreen shrubs, and tree-crowded seeps furnish shelter and nesting spaces. The climate allows a variety of rodents and insects to remain active all year, and extends the seasons of flowers and fruit. The sum of these features keeps an exuberant variety of birds washing over the shrub-steppe all year long.

Winter is the time when merlins, peregrines, and gyrfalcons join their prairie cousins in pursuit of avian prey

PRAIRIE FALCON

◄◄ AMERICAN AVOCET ◄ YELLOW WARBLER

PONDEROSA PINE IN WINTER

Winter comes to the Basin in October, as waves of hardy songbirds and raptors begin to ride in from the north. The small birds often travel in mixed flocks that can include such beauties as snow buntings, longspurs, and rosy finches. The hunters, cloaked in shades of white and dun that mirror snow on open ground, range from thrush-sized northern shrikes to round snowy owls. This is the time when merlins, peregrines, and gyrfalcons join their prairie cousins in pursuit of avian prey, jetting across the landscape after everything from featherweight horned larks to chunky ducks.

Temperature and snowpack can vary greatly through the dark months, and the shrub-steppe's windswept winters meld visions of arctic tundra with the spasmodic thaws of spring. When the cold dominates, rough-legged hawks hover against gale-force winds to track scampering voles. Resident ravens step out onto solid ice, and magpies pick their way across frozen fields. But long before winter is over, sunny days make the basalt cliffs glow with energizing heat. While visiting snowy owls stand silently in melting fields, great horned owls begin to bark out their territorial boos. By the first of March, tiny canyon wrens pipe their haunting downhill songs. Their clear notes are soon backed by the scolding flurries of the rock wren, called *Taymuusya* (news-bearer) by the Sahaptian people. White-throated swifts and violet-green swallows, their tiny bodies as boldly patterned as killer whales, arrive from far to the south to swarm around the cliffs.

Among open sagebrush and bunchgrass, robins probe the crust while single Say's phoebes, delicate in appearance but tough against the breeze, pick off the first hatches of aerial insects. Meadowlarks and half a dozen different sparrows suddenly hop up to the shrubtops to cast out their distinctive melodies. Ducks gather in tight clumps of

By the first of March, tiny canyon wrens pipe their haunting downhill songs. Their clear notes are soon backed by the scolding flurries of the rock wren, called Taymuusya *(news-bearer) by the Sahaptian people.*

ROCK WREN

SANDHILL CRANES

astonishing size, and the percussive music of tundra swans and sandhill cranes rains down from high above. Before the first week of April has passed, grouse have stamped the grasses of their dancing grounds flat.

In country known for its dryness, standing water seems to spread everywhere, and around each pond and puddle the level of intensity continues to rise. When surveyor Charles Wilson camped near the Palouse River in the spring of 1861, he found it hard to sleep amidst all the excitement: "Long after I went to bed, I lay awake listening to the wild cries of the wild fowl as they were feeding amongst the rushes." Many other early journalkeepers noted the abundance of swans, geese, ducks, shorebirds, cranes, herons, bitterns, and grebes that crowded the pothole lakes and watercourses during spring migration. Many of these waterfowl remained in the region to nest, as some of their descendants do today.

Neotropical migrants, brightly tressed and full of song, begin to filter in during the latter part of April. Most head for thick riparian growth around dependable water, where they carry on a frenzy of breeding activity. In early June, inside an oasis of trees at the head of a small coulee, the air feels packed with noise. Through the short night the concert never quite stops, as the swooping booms of evening nighthawks and the warbled notes of a poor-will overlap with the pre-dawn sneezes of a small flycatcher. Even after the sun's heat has lowered the din of song, a young spotted towhee practices scratching about for seeds beneath the shrubbery. An eastern kingbird, just two weeks past a flight that spanned most of North America, adds sticks to its new nest in an aspen crotch; she interrupts her work to spar with a robin sitting on its second clutch in the next tree. The yellow-green plumage of a female northern oriole is difficult to spot

"We saw a crane valiantly defending her nest against two 'Coyotes' or small prairie wolves—the whole thing was full in sight; the quick circles of the crane & her darts at the two wolves, who had hard work to keep beyond reach of her formidable beak, how it ended we did not see as it came on to rain heavily."

—CHARLES WILSON

"Many birds require woody plants as nesting or as perching sites, so removing Artemisia eliminates certain elements of the avifauna."

—R. Daubenmire

Mountain bluebird

► Savannah sparrow

until she sails from her bulbous hanging nest, followed by a trio of tangerine-hued fledglings. Perched right out in the sunshine, a lazuli bunting repeats its carefully phrased song over and over. Along the cliff edge above its head, four kestrels wheel and cry, old enough to hunt but not skilled enough to catch anything.

After a few breathtaking weeks, this flurry suddenly ends; much of the winged life evaporates from the sagebrush country, and the birds that remain assume quieter, more deliberate habits. In the middle of the hottest day of summer, a sage thrasher works its way through a currant bush, delicately plucking fruit. Two dozen sooty-faced lark sparrows glean a fresh burn for roasted seeds. A grasshopper freshly speared on a hopsage spine reveals the presence of a loggerhead shrike. Four western bluebirds, their trademark color obscured by browns and grays, float across the sagetops in slow motion.

Fall migration does not compare to the concentrated rush of spring, because many adult birds slip away as early as July. It is the first-year birds, with their buffy yellow plumages and unpracticed ways, that hang around, exuding the pent-up energy that will carry them south. While many shorebirds pass through quickly in August, young pectoral sandpipers arrive late and remain deep into September, uttering soft quirps as they probe the mudflats around small lakes. By then, loose clumps of dryland sparrows pinwheel silently from bush to bush, their species barely separable in their fall plumages. Water pipits wag across fallow fields as if they know exactly where they are going. Immature Swainson's hawks the color of dark chocolate hunch on power poles, watching the wheat stubble for movement. When a cold front sweeps frigid wind and a dusting of snow across the Cascades, the lingering migrants shuffle south, leaving the plains open for the panorama of winter.

"**I killed a Fowl of the Pheasent kind as large as a (Small) turkey. . . . The tail feathers 13 Inches long, feeds on grass hoppers, and the Seed of wild Isoop [Big sagebrush].**"

—MERIWETHER LEWIS

SAGE GROUSE ON DANCING GROUNDS

During Charles Wilson's trip through the scablands in the late summer of 1861, he "had fine fun racing & galloping about on the Prairies, occasionally stopping to have a shot at a curlew or grouse that started up at our approach." The birds that Wilson was picking off have all suffered sharp declines in the years since he passed through, and two species of grouse illustrate this common theme.

The strong flight and ethereal dances of both sage and sharp-tailed grouse figure strongly in the stories of several Plateau tribes, as well as in the accounts of Basin homesteaders who regularly enjoyed the tasty birds for supper. For ten months of the year, football-shaped droppings and an occasional lost feather are the closest most people get to these birds; even when flushed, their dappled plumage seems to keep them attached to the earth. Yet by the Ides of March, their neck and tail feathers begin to prickle out like spears, and glands on both sides of their necks and under their wings bulge with shocking colors. Birds from the far reaches of their territory gather at breeding sites (called leks) atop small knolls in open country. Males and females remain

COLUMBIA SHARP-TAILED GROUSE

"In certain places they are in great numbers in the autumn, congregating in large flocks, especially in the vicinity of patches of wild rye, and more recently near settlements where there are wheat stubbles."

—J.G. COOPER

around the leks for some weeks as they strut through their measured, fluent, wildly ecstatic ritual of courtship.

Lewis and Clark called the sage grouse "the cocks of the plains" for their strutting habits and short stiff neck feathers. These birds depend entirely on sagebrush for food, shelter, nesting sites, and wintering grounds. As large tracts of sage have been broken up, their numbers in eastern Washington have fallen to a few hundred individuals. For a bird of such bulk, they can be very secretive, and their distribution remains somewhat obscure.

Sharp-tailed grouse are more a bird of the bunchgrass, but they also need protected nesting sites, good hatches of insects to feed their young, grasshoppers and fruit in late summer and fall, and budding aspen trees along watercourses in the winter. For a few decades in the late 1800s, the stubble of new grainfields provided a rich new food source for sharp-taileds, and their numbers increased. But as more and more acreage was given over to livestock grazing or lost to the plow, the diversity of habitat that the birds require began to disappear. Farm women were no longer able to send their children

"**I am inclined to think that their abodes in this vicinity are among the fissures and cracks of the basaltic rock so abundant there, rather than in the soft earth.**"

—**J.G. Cooper**

Young burrowing owls

out with gunny sacks and a .22 to procure an evening meal of prairie hens. When the practice of burning spring wheat stubble took hold soon after the turn of the century, sharp-tailed populations plummeted.

Reduced grouse hunting seasons in 1920 and strict closures after 1933 provided little relief, and by 1955 the birds had disappeared from six eastern Washington counties. In 1970, the Game Department estimated a statewide population of around seventy-five hundred birds scattered in only three counties, with the most stable population on the Colville Indian Reservation. Since then, regular lek surveys show that attempts at protection have not halted that downward spiral; a 1997 census counted fewer than one thousand birds.

A century and a half of written records present plenty of other examples of such alarming decline. From sage thrashers to black terns, from loggerhead shrikes to ferruginous hawks, over thirty birds that frequent the shrub-steppe are now monitored as species of concern. Yet certain of them show remarkable perseverance and adaptability. Trumpeter swans were given up as a lost species thirty years ago, but for the past few springs a small group of them has stopped at a seasonal pond near Harrington to rest and refuel on emerging wheat. White pelicans put on a long slide toward oblivion, then partially rebounded; now increasing numbers summer on a few permanent lakes around the Potholes. Blue-legged avocets and black-necked stilts keep expanding their range around waste ponds and irrigated waterways. The explosive grunts of the yellow-breasted chat still rise from springs where cows have trampled all the native plants away. Most burrowing owls were driven from the steppe as plows collapsed their burrows, but some pairs have improvised nesting sites in irrigation culverts. Their yellow eyes stare out, patiently waiting, as tractors rumble past.

AMERICAN WHITE PELICANS

beyond what we can see

THE FUTURE OF THE SHRUB-STEPPE

In mid-October of 1805, Meriwether Lewis and William Clark were working their way through the basalt canyons of the Snake River, closing in on their historic meeting with the Columbia. One day upstream from the confluence, much to their annoyance, they encountered "great quantities of a kind of prickley pares." Although Lewis and Clark had battled this cactus that clings tenaciously to men and horses on the buffalo plains of Montana, the ones along the Snake were "much worst than any I have before Seen," Clark wrote, "of a tapering form, and attach themselves by bunches."

Prickly pear cacti, notable for their lemon- and peach-colored flowers as well as their nettlesome spines, usually occur as scattered individual plants. But any segment

"**How much longer we might have been tempted to stay here it is impossible to say, but we accidentally set fire to the grass.**"

—PAUL KANE

SPROUTING ELDERBERRY AFTER FIRE

◄◄ SHRUB-STEPPE AND FARMLAND IN THE KITTITAS VALLEY ◄ COLUMBIA RIVER KAYAKERS ALONG HANFORD REACH

that is torn off a mother cactus can reroot wherever it lands, and the more the prickly pears are broken and scattered, the denser the carpet they form. In 1805 there were no buffalo or cattle at the confluence of the Snake and the Columbia, but it was a center of horse country, where surrounding tribes had been breeding and trading animals since they first obtained them almost a century before. It seems likely that the quantities of prickly pears were the result of trampling by horses, and that Clark was bearing witness to a small crescent of change that crept along the river. His own expedition presaged alterations on a much greater scale, and the mechanization of the twentieth century accelerated many of them to breakneck speed.

Today, the view from an airplane window reveals a Great Columbia Plain so different from the one visited by Lewis and Clark that it seems to have been swept by a new geologic age. Dams have swollen the sinuous ribbon of the Columbia River into a series of placid lakes that harbor introduced fish and Eurasian snails adapted to warm, turbid water. Concrete canals stretch like spider legs from those storage lakes, ferrying acre-feet of water to places that had been dry since the Lake Missoula Floods. Grain farms create patchworks of light green ground cover and fallow brown soil. The darker green circles of center pivots, many fed from deep, aquifer-tapping wells, form a dot pattern of prosperous agricultural country. New apple orchards march across hillsides in orderly rows, and any stand of sagebrush within easy reach of water has almost certainly disappeared.

Yet inside this scheme of transported color and moisture, a few areas still retain the luminous gray-green glow of healthy shrub-steppe. Both in and out of the region, there is a growing awareness that something precious still exists here, that the grand expanses of

State Highway 261 at Lyons Ferry

"Man's ingenuity now promises the Columbia Plateau an optimum of water, delivered through the long cloudless summers. Wheat farming will give place . . . to alfalfa growing and fruit raising. In the unclaimed or abandoned dry gravel soils sagebrush and bunch grass will vanish, and orchards will spread a summer greenery over the gray landscape."

—J Harlen Bretz

APPLE ORCHARD

sagebrush and bunchgrass hold real value, in the fullest ecological and cultural senses. As increasing numbers of retirees and urban refugees move to the rural reaches of eastern Washington, the future of these valuable remnants has become a question of paramount importance. A few pieces have been protected by means such as the Arid Lands Ecology Reserve at Hanford, national wildlife refuges, and Nature Conservancy preserves. Other viable tracts can be found under a variety of public, private, and tribal management.

Right now, many of these parcels have arrived at a moment of transition. Economic pressures are driving families away from small ranches, which are then absorbed by agribusiness or broken up for development. The Department of Energy is preparing to

AERIAL OF ALFALFA HARVEST

"**For several days we saw only great sleepy-looking hills, stretching in endless succession, as far as the horizon extended, from morning till night, as if a billowy ocean had been suddenly transfixed in the midst of its motion.**"

—Caroline C. Leighton

spin off large portions of the Hanford Site, and several different interests are vying for control. At the same time, the roles of government agencies like the Bureau of Land Management and the Department of Fish and Wildlife are being reevaluated in ways that will affect large areas of the Plateau. Together, these factors will soon change the pattern of ownership within the region.

For any of the true shrub-steppe to survive, these new stewards will have to balance the benefits of development with the value of wild, open space. Whoever touches this land next must also take the full sweep of the shrub-steppe into account. It is possible to look at its mosaic now, from the most scenic bluffs to the ashiest, most despondent burn, and see it as the essential base upon which a workable future can be built. Despite the thorny problems that face this region, there is much here to celebrate. There are still places where pink bitterroot flowers crawl across cracked basalt, just as they have since the waning of the last Ice Age. Prairie falcons stoop on migrating shore-birds, and wood rats carry their seeds to ancient middens. Along favored hillsides, for a few days each year, groups of women still prod the ground with digging sticks, coaxing out their favorite roots; the walls of isolated coulees still echo with the sound of fast water tumbling stones. The strains of these remembered rhythms resonate through the landscape, drawing strength from the whole scope of its human and natural history.

Saddle Mountains, Wahluke Slope ▶ Sumac in hills surrounding Yakima Valley

further reading on the shrub-steppe

EUROPEAN EXPLORATION AND SETTLEMENT

Forgotten Trails: Historical Sources of the Columbia's Big Bend Country. Ron Anglin. Washington State University Press, Pullman, 1995.

Journals of Lewis & Clark. Gary Moulton, ed. Vols. 5 & 7, University of Nebraska Press, Lincoln, 1988 & 1991.

Paul Kane's Frontier. Russell Harper, ed. Amon Carter Museum and University of Texas Press, Austin, 1971.

Sagebrush Homesteads. Laura Tice Lage. Franklin Press, Yakima, Washington, 1967.

Sources of the River: Tracking David Thompson across Western North America. Jack Nisbet. Sasquatch Books, Seattle, Washington, 1994.

West Coast Journeys 1865-1879. Caroline C. Leighton, introduced by David Buerge. Sasquatch Books, Seattle, Washington, 1995.

FAUNA

Amphibians of Washington and Oregon. Leonard, Brown, Jones, McAllister, & Storm. Seattle Audubon Society, Seattle, Washington, 1993.

Before the Indians. Bjorn Kurten. Columbia University Press, New York, 1988.

Birds of the Channeled Scablands of Eastern Washington. Bureau of Land Management pamphlet, 1995.

Birds of Yakima Canyon. Bureau of Land Management pamphlet.

The Hanford Reach: What do We Stand to Lose? David R. Geist, *Illahee*, pp. 130-141, Vol. 11, Nos. 3 & 4, 1995.

I am of this Land, Wetes pe m'e wes: Animals of Hanford Site, A Nez Perce Nature Guide. Comp. by Dan Landeen and Jeremey Crow. Western Printing, Lewiston, Idaho, 1997.

Reptiles of Washington and Oregon. Brown, Bury, Darda, Diller, Peterson, & Storm. Seattle Audubon Society, Seattle, Washington, 1995.

Salmon Fishers of the Columbia. Courtland L. Smith. Oregon State University Press, Corvallis, 1979.

FLORA

After the Ice Age: The Return of Life to Glaciated North America. E.C. Pielou. University of Chicago Press, Chicago, 1991.

Food Plants of Interior First Peoples. Nancy J. Turner. University of British Columbia Press, Vancouver, 1997.

Northwest Weeds. Ron Taylor. Mountain Press Publishing, Missoula, Montana, 1990.

Sagebrush Country: A Wildflower Sanctuary. Ronald J. Taylor. Mountain Press Publishing, Missoula, Montana, 1990.

Steppe Vegetation of Washington. F. R. Daubenmire. Washington Agricultural Experiment Station Technical Bulletin 62, Pullman, 1970.

Trees, Shrubs, and Flowers of British Columbia and Washington State. C.P. Lyons, Lone Pine Press, Vancouver, British Columbia, 1994.

Watching Washington Wildflowers. Bureau of Land Management pamphlet.

GEOLOGY

Cataclysms on the Columbia. John Eliot Allen and Marjorie Burns. Timber Press, Portland, Oregon, 1986.

The Channeled Scablands. Victor Baker and Dag Nummedal. Planetary Geology Program, Office of Space Science, NASA, Washington, D.C., 1978.

Fire, Faults, & Floods: A Road & Trail Guide Exploring the Origins of the Columbia River Basin. Marge & Ted Mueller. University of Idaho Press, Moscow, 1997.

Ginko Petrified Forest. Mark Orsen. Ginkgo Gem Shop, Vantage, Washington.

The Grand Coulee. J Harlen Bretz. Special Publication No. 15, American Geographical Society, New York, 1932.

Northwest Exposures: A Geologic Story of the Northwest. David Alt and Donald Hyndman. Mountain Press, Missoula, Montana, 1995.

TRIBES

Ghost Voices: Yakima Indian Myths, Legends, Humor, and Hunting Stories. Donald Hines. Great Eagle Publishing Co., Issaquah, Washington, 1992.

Half-Sun on the Columbia: A Biography of Chief Moses. Robert Ruby and John Brown. University of Oklahoma Press, Norman, 1965.

Mourning Dove: A Salishan Autobiography. University of Nebraska Press, Lincoln, 1990.

Nch'i-Wana, The Big River: Mid-Coumbia Indians and Their Land. Eugene S. Hunn with James Selam and Family. University of Washington, Seattle, 1990.

The Richey Clovis Cache: Earliest Americans along the Columbia River. Richard Michael Gramly. Persimmon Press, Kenmore, Washington, 1993.

photo credits/copyright

index

PHOTO CREDITS

Bertelson, Ernest (Special Collections Division, University of Washington Libraries), 34
Bowman, L. E., 61, 117
Courtesy Burke Museum of Natural History and Culture, 33, 36
Curtis, Asahel (Special Collections Division, University of Washington Libraries), 25, 39
Curtis, Asahel (Washington State Historical Society, Tacoma), 37, 38
Curtis, Edward S. (Special Collections Division, University of Washington Libraries), 22
Dunwiddie, Peter (The Nature Conservancy), 53
Elk, John, III, 75 left
Fobes, Natalie, 34, 68, 69, 71, 72, 73, 74
Keeley, Ernest R., 70, 75 right, 76, 77
Keyser, James D., 32
Landers, Rich, 105
Lazelle, Keith, cover, 1, 4, 5, 6, 7, 8, 48, 49, 57, 58, 60, 62, 83, 89, 93, 96, 97, 99, 102, 103, 104, 107, 115, 118
Leeson, Tom and Pat, 86
Leonard, William, 63, 66, 78, 79, 80, 81, 82, 85
Marshall, John, 3, 10, 13, 16, 17, 19, 28, 30, 41, 42, 47, 51, 55, 56, 65, 91, 98 right, 108, 111, 112, 116, 120
O'Hara, Pat, 14
Rogers, Joel, 9, 20, 67, 109
Sheehan, Mark, 2, 18, 27, 31, 54, 87, 110
Special Collections Division, University of Washington Libraries, 35
Wells, Randy, 45, 113
Wolfe, Art, 98 left, 100, 106

Library of Congress Cataloging-in-Publication Data:

Nisbet, Jack, 1949-
    Singing grass, burning sage : discovering Washington's shrub-steppe / text by Jack Nisbet.
        p. cm.
    "A Nature Conservancy of Washington book."
    ISBN 1-55868-478-6
    1. Natural history—Washington (State) 2. Steppe ecology—Washington (State) 3. Natural History—Washington (State) Pictoral works. I. Title.
    QH105.W2 N57   1999
    508.797—dc21   99-16619
                                                        CIP

Design: Elizabeth Watson
Map type: Michelle Taverniti
Printed in Singapore
Second Printing

Co-published by
    The Nature Conservancy of Washington
    217 Pine Street, Suite 1100
    Seattle, Washington 98101
    206/343-4344
    www.tnc-washington.org

                and

    Graphic Arts Center Publishing®
    An imprint of Graphic Arts Center Publishing Company
    P.O. Box 10306
    Portland, Oregon 97296-0306
    503/226-2402
    www.gacpc.com